Peak Performance
How to Achieve and Sustain Excellence in Operations Management

Sustainable Improvements in Environment Safety and Health

Series Editor: Frances Alston, Lawrence Livermore National Laboratory, California, USA

Lean Implementation
Applications and Hidden Costs
Frances Alston

Safety Culture and High-Risk Environments
A Leadership Perspective
Cindy L. Caldwell

Industrial Hygiene
Improving Worker Health through an Operational Risk Approach
Frances Alston, Emily J. Millikin, and Willie Piispanen

The Legal Aspects of Industrial Hygiene and Safety
Kurt W. Dreger

Peak Performance: How to Achieve and Sustain Excellence in Operations Management
Patricia M. Allen, Frances E. Alston, and Emily Millikin DeKerchove

For more information about this series, please visit: https://www.crcpress.com/Sustainable-Improvements-in-Environment-Safety-and-Health/book-series/CRCSUSIMPENVSAF

Peak Performance
How to Achieve and Sustain Excellence in Operations Management

Authored by
Patricia Melton Allen, Frances E. Alston, and Emily Millikin DeKerchove

CRC Press
Taylor & Francis Group
Boca Raton London New York

CRC Press is an imprint of the
Taylor & Francis Group, an **informa** business

CRC Press
Taylor & Francis Group
6000 Broken Sound Parkway NW, Suite 300
Boca Raton, FL 33487-2742

First issued in paperback 2021

© 2019 by Taylor & Francis Group, LLC
CRC Press is an imprint of Taylor & Francis Group, an Informa business

No claim to original U.S. Government works

ISBN-13: 978-0-367-77979-5 (pbk)
ISBN-13: 978-1-138-32324-7 (hbk)

This book contains information obtained from authentic and highly regarded sources. Reasonable efforts have been made to publish reliable data and information, but the author and publisher cannot assume responsibility for the validity of all materials or the consequences of their use. The authors and publishers have attempted to trace the copyright holders of all material reproduced in this publication and apologize to copyright holders if permission to publish in this form has not been obtained. If any copyright material has not been acknowledged, please write and let us know so we may rectify in any future reprint.

Except as permitted under U.S. Copyright Law, no part of this book may be reprinted, reproduced, transmitted, or utilized in any form by any electronic, mechanical, or other means, now known or hereafter invented, including photocopying, microfilming, and recording, or in any information storage or retrieval system, without written permission from the publishers.

For permission to photocopy or use material electronically from this work, please access www.copyright.com (http://www.copyright.com/) or contact the Copyright Clearance Center, Inc. (CCC), 222 Rosewood Drive, Danvers, MA 01923, 978-750-8400. CCC is a not-for-profit organization that provides licenses and registration for a variety of users. For organizations that have been granted a photocopy license by the CCC, a separate system of payment has been arranged.

Trademark Notice: Product or corporate names may be trademarks or registered trademarks, and are used only for identification and explanation without intent to infringe.

Library of Congress Cataloging-in-Publication Data

Names: Allen, Patricia M., author. | Alston, Frances (Industrial engineer), author. | DeKerchove, Emily Millikin, author.
Title: Peak performance : how to achieve and sustain excellence in operations management / authored by Patricia Melton Allen, Frances E. Alston, and Emily Millikin DeKerchove.
Description: Boca Raton : CRC Press, Taylor & Francis, 2019. | Series: Sustainable improvements in environment safety and health | Includes bibliographical references and index.
Identifiers: LCCN 2019001291| ISBN 9781138323247 (hardback : alk. paper) | ISBN 9780429451508 (e-book)
Subjects: LCSH: Organizational effectiveness. | Personnel management. | Operations research.
Classification: LCC HD58.9 .A455 2019 | DDC 658.5/03--dc23
LC record available at https://lccn.loc.gov/2019001291

Visit the Taylor & Francis Web site at
http://www.taylorandfrancis.com

and the CRC Press Web site at
http://www.crcpress.com

This book is dedicated to our mothers: Mildred Y. Melton, Ella Singleton, and Patricia J. Griffin

Contents

Preface xv
Authors xix

Chapter 1 Defining peak performance and why it matters 1
1.1 Introduction 1
1.2 Strategic management 2
1.3 Workplace dynamics 3
1.4 Peak performance model 4
    1.4.1 Lean 4
    1.4.2 Human performance 6
        1.4.2.1 Human behavior and emotions 6
    1.4.3 Operations excellence 7
    1.4.4 Organization culture 8
1.5 The impact of behaviors on organization performance 10
    1.5.1 What do we mean when we refer to human behavior? 11
    1.5.2 Leadership behavior and style 11
1.6 Organizational performance 13
1.7 The peak performance model 13
References 14

Chapter 2 Principles of lean to optimize business outcomes 15
2.1 Introduction 15
2.2 Why lean? 16
2.3 Evaluating the waste in the process 17
    2.3.1 Travel time 17
    2.3.2 Inventory 19
    2.3.3 Excess motion 19
    2.3.4 Waiting periods 21
    2.3.5 Overproduction 21
    2.3.6 Over-processing 22
    2.3.7 Rework 22
    2.3.8 Human capital 23

2.4 Lean tools 24
　2.4.1 The five action-steps to lean 25
　　2.4.1.1 Step 1 – Identify the problem 25
　　2.4.1.2 Step 2 – Measure the targeted work process 29
　　2.4.1.3 Step 3 – Analyze the data 31
　　2.4.1.4 Step 4 – Improve the targeted work process 34
　　2.4.1.5 Step 5 – Sustain process improvement 37
2.5 Assessing lean principles 38
References 40

Chapter 3 Principles of human performance and performance modes 41
3.1 Introduction 41
3.2 People are fallible 41
　3.2.1 Active and latent errors 43
3.3 Anatomy of an event 44
　3.3.1 Error precursors 44
　　3.3.1.1 Task requirement 46
　　3.3.1.2 Employee competencies 46
　　3.3.1.3 Work settings 47
　　3.3.1.4 Human characteristics 48
　3.3.2 Initiating event 48
　3.3.3 Latent organizational weakness 49
　3.3.4 Defective controls 49
3.4 Organizational culture 50
3.5 Human performance modes 51
　3.5.1 Skill-based performance mode 52
　3.5.2 Rule-based performance mode 54
　3.5.3 Knowledge-based performance mode 58
3.6 Assessing Human Performance Principles 62
References 63

Chapter 4 Human error defenses 65
4.1 Introduction 65
4.2 Human error prevention 65
　4.2.1 Task preview 66
　　4.2.1.1 When to utilize the tool 67
　　4.2.1.2 At-risk practices to consider preventing 67
　4.2.2 Job-site review 67
　　4.2.2.1 When to utilize the tool 68
　　4.2.2.2 At-risk practices to consider preventing 68
　4.2.3 Questioning attitude 68
　　4.2.3.1 When to utilize the tool 69
　　4.2.3.2 At-risk practices to consider preventing 69

Contents

	4.2.4	Pause when unsure or uncertain	70
		4.2.4.1 When to utilize the tool	70
		4.2.4.2 At-risk practices to consider preventing	70
	4.2.5	Self-checking	70
		4.2.5.1 When to utilize the tool	71
		4.2.5.2 At-risk practices to consider preventing	71
	4.2.6	Procedure use and compliance	72
		4.2.6.1 When to utilize the tool	74
		4.2.6.2 At-risk practices to consider preventing	74
	4.2.7	Effective communication	74
		4.2.7.1 Three-way communications	75
		4.2.7.2 Phonetic alphabet	76
	4.2.8	Place-keeping	78
		4.2.8.1 When to utilize the tool	78
		4.2.8.2 At-risk practices to consider preventing	78
	4.2.9	Confirming assumptions	78
		4.2.9.1 When to utilize the tool	79
		4.2.9.2 At-Risk practices to consider preventing	79
4.3	Human-error prevention techniques for teams	79
	4.3.1	Peer-checking	83
		4.3.1.1 When to utilize the tool	84
		4.3.1.2 At-risk practices to consider preventing	84
	4.3.2	Independent verification	84
		4.3.2.1 When to utilize the tool	85
		4.3.2.2 At-risk practices to consider preventing	85
	4.3.3	Flagging	85
		4.3.3.1 When to utilize the tool	86
		4.3.3.2 At-risk practices to consider preventing	86
	4.3.4	Turnover briefing	87
		4.3.4.1 When to utilize the tool	87
		4.3.4.2 At-risk practices to consider preventing	88
	4.3.5	Post-job review	88
		4.3.5.1 When to utilize the tool	89
		4.3.5.2 At-risk practices to consider preventing	89
	4.3.6	Project planning	89
		4.3.6.1 When to utilize the tool	90
		4.3.6.2 At-risk practices to consider preventing	90
	4.3.7	Problem-solving	90
		4.3.7.1 When to utilize the tool	91
		4.3.7.2 At-risk practices to consider preventing	91
	4.3.8	Decision-making	91
		4.3.8.1 When to utilize the tool	92
		4.3.8.2 At-risk practices to consider preventing	93

	4.3.9	Contractor oversight	93
		4.3.9.1 When to utilize the tool	93
		4.3.9.2 At-risk practices to consider preventing	94
4.4	Human error prevention techniques for managers		94
	4.4.1	Benchmarking the competition and industry leaders	94
		4.4.1.1 When to utilize the tool	95
		4.4.1.2 At-risk practices to consider preventing	95
	4.4.2	Observation of work and behaviors	96
		4.4.2.1 When to utilize the tool	96
		4.4.2.2 At-risk practices to consider preventing	96
	4.4.3	Self-assessments	97
		4.4.3.1 When to utilize the tool	97
		4.4.3.2 At-risk practices to consider preventing	97
	4.4.4	Performance indicators and application	98
		4.4.4.1 When to utilize the tool	99
		4.4.4.2 At-risk practices to consider preventing	99
	4.4.5	Independent oversight	99
		4.4.5.1 When to utilize the tool	100
		4.4.5.2 At-risk practices to consider preventing	100
	4.4.6	Work product evaluation	101
		4.4.6.1 When to utilize the tool	101
		4.4.6.2 At-risk practices to consider preventing	101
	4.4.7	Change management	102
		4.4.7.1 When to utilize the tool	102
		4.4.7.2 At-risk practices to consider preventing	102
	4.4.8	Reporting errors and near misses	102
		4.4.8.1 When to utilize the tool	103
		4.4.8.2 At-risk practices to consider preventing	103
4.5	Critical steps		103
	4.5.1	Task preview	104
	4.5.2	Pre-job briefing	104
	4.5.3	Procedure use	105
	4.5.4	How to get started using critical steps	106
References			108

Chapter 5 Operations excellence 109
5.1 Introduction 109
    5.1.1 Core function one: Identify mission-critical processes 111
    5.1.2 Core function two: Identify mission-critical operations and activities within each mission-critical process 112
    5.1.3 Core function three: Establish Operations Excellence Program Elements 113
        5.1.3.1 Organization and administration 113

		5.1.3.2	Communications..114
		5.1.3.3	Training programs..114
		5.1.3.4	Equipment configuration management status and control...114
	5.1.4	Core function four: Implement the Operations Excellence Program 117	
	5.1.5	Core function five: Evaluate and implement improvements..119	
5.2	Application of an Operations Excellence Program......................... 121		
	5.2.1	Core function one: Identify operations excellence mission-critical processes at the Ruby Rabbit Company...... 121	
	5.2.2	Core function two: Identify critical operations and activities within each mission-critical process at the Ruby Rabbit Company 123	
	5.2.3	Core three: Establish Operations Excellence Program Elements for the Ruby Rabbit Company 126	
	5.2.4	Core function four: Implement the Ruby Rabbit Operations Excellence Program 126	
	5.2.5	Core function five: Evaluate and implement improvements at the Ruby Rabbit Company..................... 128	

Chapter 6 Culture: The bedrock for sustainable organizational performance 129

6.1 Introduction.. 129
6.2 Whistleblower 130
6.3 Elements and attributes of culture for peak performance 132
    6.3.1 Trusting cultures 132
    6.3.2 Learning culture 133
    6.3.3 Safety culture 134
6.4 Role of leadership in culture development and sustainment 135
    6.4.1 Followership in workplace cultures.................................... 137
6.5 Evaluating the health of organization culture 139
    6.5.1 Policy and procedure reviews 139
    6.5.2 Focus group and individual interviews.............................. 145
    6.5.3 Culture sustainability strategy 145
References.. 146

Chapter 7 The peak performance model...................................... 147

7.1 Introduction.. 147
7.2 Lean 149
7.3 Human performance improvement.. 150
7.4 Operations excellence 150
7.5 Organizational culture 151
7.6 Integration of peak performance elements 151

7.7	Application of the peak performance model		152
	7.7.1	Quantitative and qualitative evaluation of peak performance model functional elements	154
	7.7.2	Evaluation of peak performance model functional elements	155
	7.7.3	Collective evaluation of peak performance model	156
7.8	Development of a peak performance improvement plan		156
7.9	Journey to achieving and sustaining peak performance		158

Chapter 8 Practitioner's guide case study: Spud's Chemical Company, LLC 161

Chapter 9 The peak performance model applied to Spud's Chemical Company, LLC 169

9.1	Introduction			169
9.2	Qualitative evaluation of Spud's Chemical Company			169
	9.2.1	Lean program activities		170
		9.2.1.1	Lean summary	174
	9.2.2	Human Performance Improvement (HPI)		175
		9.2.2.1	Human Performance Improvement (HPI) summary	181
	9.2.3	Operations excellence		182
		9.2.3.1	Document reviews	182
		9.2.3.2	Interview questions and responses	182
		9.2.3.3	Operations excellence summary	187
	9.2.4	Organizational culture		188
		9.2.4.1	Document review	189
		9.2.4.2	Employee engagement survey	189
		9.2.4.3	Trust survey	190
		9.2.4.4	Organizational culture survey	190
		9.2.4.5	Organization culture focus-group interview	192
		9.2.4.6	Organizational culture individual interviews	193
		9.2.4.7	Organizational culture summary	194
9.3	Quantitative evaluation of Spud's Chemical Company			195
	9.3.1	Lean		195
		9.3.1.1	Leadership	197
		9.3.1.2	Employee engagement	197
		9.3.1.3	Organizational factors	197
		9.3.1.4	Work planning and execution	197
	9.3.2	Human performance improvement		197
		9.3.2.1	Leadership	198
		9.3.2.2	Employee engagement	199
		9.3.2.3	Organizational factors	199
		9.2.2.4	Work planning and execution	199

	9.3.3	Operations excellence	199
		9.3.3.1 Leadership	200
		9.3.3.2 Employee engagement	201
		9.3.3.3 Organizational factors	201
		9.3.3.4 Work planning and execution	201
	9.3.4	Organizational culture	201
		9.3.4.1 Leadership	202
		9.3.4.2 Employee engagement	203
		9.3.4.3 Organizational factors	203
		9.3.4.4 Work planning and execution	203
		9.3.2.5 Cumulative quantitative analysis	203
9.4	Spud's Chemical Company Operational Improvement Plan		206

Index 211

Preface

Why is it that some companies have outstanding production rates, produce quality products, and experience record sales, with minimal injuries? These companies appear to have similar operations and employees, and similar organizational structures. Is the equipment they use better? Are the employees more knowledgeable and better skilled? Is the leadership team more skilled? Perfection and performance are key concepts that corporate leaders have been wrestling with for decades in order to continue to improve business outcomes. The reality is that perfection is often desired, but rarely achieved. In fact, we all strive for perfection in our daily lives. The problem with being able to achieve perfection in our daily lives as well as in business is that it involves the actions and activities of humans who are fallible. It is recognized that human error is normal and a natural part of being a human being; mistakes will be made.

High-performing businesses require their employees to perform their work activities safely and in compliance with requirements, as well as continuously seeking ways to improve the process/operation. There are four programs, which many industries use to achieve higher performance:

- Lean thinking
- Human Performance Improvement
- Operations Excellence
- Organizational Culture

The key to success and reaching Peak Performance is integrating three processes to optimize your business – Lean, Human Performance Improvement, Operations Excellence – while exhibiting a strong organizational culture as the underpinning. Collectively, these four components form the basis of the Peak Performance Model.

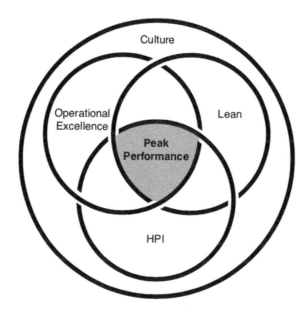

What is Peak Performance for an organization? It is when the organization has implemented streamlined processes, reduced waste and rework, human error is anticipated and managed to reduce process upsets, employees follow policies and procedures, all unpinned with a culture of trust and continuous improvement.

Various components of this model are used in industries as stand-alone programs or process improvement practices. Lean is a way of thinking that leads to instituting process and practices that create value for the customer through the elimination of waste. System and process waste has a tendency to create additional cost for a company that is generally passed on to their customers. Lean is key in helping organizations increase efficiency and provide the customer value. Implementing Lean requires an understanding of how company or operational processes work, a Lean-thinking leadership team, and engaged employees that are able to think outside of the box. All across the United States, companies are using the Lean Process, Six Sigma, or a combination of Lean Six Sigma to target and identify unnecessary waste, or tasks being performed, which add no value to the customer and improve business efficiency (along with profits).

Multiple studies have shown that the prevalent source of incidents and events are attributed to human error. The cost of events triggered by human error is undeniable. Other studies have shown that managing human error throughout the life cycle of a project will yield a positive return on investment. Depending on the reports and studies selected, human beings have error rates ranging anywhere from 1 to 15 errors per hour. One study asserts that people, when performing any activity, commit on average three to four errors per hour. Typical human reliability is on the order of 99.99% (1 error out of 1000 opportunities), depending on the task and the immediate working conditions.

Operations Excellence is a program implemented by companies or industries that desire proficiency in performing essential and routine tasks in the workplace. Often Operations Excellence programs are found in industries that require a high rate of proficiency in task performance, such as the nuclear industry, space industry, medical field, and military forces. These industries apply an Operations Excellence Program to instill discipline and rigor on key work tasks that are critical to successful operations. In many industries an Operations Excellence Program may contain 18 functional elements, critical to the operation and mission, which drive consistency in performance; however, as described in this book, functional elements of an Operations Excellence Program will vary dependent upon mission-critical operations of the business.

The culture of an organization serves as the bedrock upon which members of the organization interact, form relationships, and support business processes and practices. When referring to organizational culture one is referring to attributes of the management systems that include practices, values, beliefs, behaviors, languages, and symbols. These attributes serve as a bedrock for the cultural health of the work environment to ensure that Peak Performance can be obtained and sustained. In addition, cultural attributes can be a great facilitator to relationship-building and instilling trust in organizations. Recognizing the fallibility of human actions, implementing Operations Excellence supported by a dynamic culture, and providing clear expectations for continuous improvement (Lean) are necessary to significantly reduce human error and reach Peak Performance.

Achieving Peak Performance requires a dynamic culture that serves as the underpinning for the organization, promotes excellence in operations, drives continuous improvement of the business, and targets employee engagement and performance. This may seem like an easy formula, but why is it so difficult to achieve? The main reason is that most organizations are focused on one or the other of these areas, but not all four at the same time, and, because they do not have a process or vision to integrate them, they are missing an opportunity to evolve into a high-performing organization. So, can we really expect to reach perfection or Peak Performance in organizations? The Peak Performance Model offers a process improvement method that is designed to help organizations achieve high performance that is sustainable long term.

There are a variety of tools that can assist an organization in achieving Peak Performance if these tools are used properly and at the right time. Until now, there has not been a concerted effort to develop and present a philosophy and concepts that Managers can use to evaluate their organizations to determine whether the organization is functioning at its optimal level. This book strategically defines elements of a model that can be used by any company or organization to enhance operational performance, process efficiencies, and improve customer satisfaction. In effect, it allows an organization to take small bites into their improvement efforts on the way to achieving optimal performance and reaching Peak Performance of the organization.

The tools presented in this book, along with an example case study, demonstrates how the model can be applied and integrated into any company's practices. The key is making assessments to determine the level of implementation, and then improving how work is performed. Every company or organization wants to improve their performance, but not many have integrated the right tools and process to achieve the high level of performance they seek. Through application of this book, every company can begin their journey to achieve and sustain Peak Performance!

Authors

Patricia Melton Allen has over 35 years' experience in Engineering, Operations, Environment, Safety and Health, Quality Assurance, Contractor Assurance, and Human Performance. Ms. Allen has held various executive leadership positions in complex operations, leading large organizations in the development and implementation of integrated safety management, environmental compliance, radiological protection, industrial safety, industrial hygiene, training and procedures, quality assurance, and emergency preparedness programs.

Ms. Allen has led organizations to achieve significant cost saving by implementing Lean tools, optimizing operations, and creatively expanding the skill sets of employees to accomplish more work. Ms. Allen has a Bachelor of Science from the University of Louisville and a Master of Science in Chemical Engineering from the University of Louisville. She holds a certification as Safety Trained Supervisor, Certified Performance Technologist, and is qualified as a Lead NQA-1 Auditor. Ms. Allen holds a patent associated with a Mobile Slip Simulator. She is a member of the American Institute of Chemical Engineers, American Society of Safety Professional, International Society for Performance Improvement, and the National Management Association, and serves on the Board of Directors for the Central Savannah River chapter of the National Management Association.

Dr. Frances E. Alston has built a solid career foundation over the past 30 years in leading the development and management of Environment, Safety, Health, and Quality (ESH&Q) programs in diverse cultural environments. Throughout her career, she has delivered superior performance within complex, multi-stakeholder situations and has effectively dealt with challenging safety, operational, programmatic, regulatory, and environmental issues.

Dr. Alston has been effective in facilitating integration of ESH&Q programs and policies as a core business function while leading a staff of business, scientific, and technical professionals. She is skilled in providing technical expertise in regulatory and compliance arenas as well as determining necessary and sufficient program requirements to ensure employee and public safety, including environmental stewardship and sustainability. Dr. Alston also has extensive knowledge and experience in assessing programs and cultures to determine areas for improvement and development of strategy for improvement.

Dr. Alston holds a B.S. degree in Industrial Hygiene and Safety, an M.S. degree in Hazardous and Water Materials Management/Environmental Engineering, an M.S.E. in Systems Engineering/Engineering Management, and a Ph.D. in Industrial and Systems Engineering.

Dr. Alston is a fellow of the American Society for Engineering Management (ASEM) and holds certifications as a certified hazardous materials manager (CHMM) and a professional engineering manager (PEM). She is an adjutant professor and the 2018 president for the American Society for Engineering Manager.

Emily Millikin DeKerchove has over 33 years of leadership experience in regulatory, environmental, radiation protection, and safety and health at Department of Energy (DOE) and Department of Defense (DOD) chemically and radiologically contaminated sites. She has served as the Director of Safety, Health, and Quality programs and managed all aspects of program, cost, and field implementation of Safety and Health, Industrial Hygiene, Radiological Control, Quality Assurance, Contractor Assurance System, Emergency Preparedness, Safeguards and Security, Occupational Health, and Price-Anderson Amendment Act Programs. Additionally, she has established and successfully led employee behavioral-based safety-observation programs and successfully achieved Voluntary Protection Program (VPP) Star status.

Ms. Millikin earned a B.S. in Environmental Health with double majors in Industrial Hygiene and Health Physics from Purdue University. She is a Certified Safety Professional and Certified Industrial Hygienist.

chapter one

Defining peak performance and why it matters

1.1 Introduction

In today's competitive business environment, many companies struggle to survive and realize a profit for their investors. The leaders of these companies have attempted to address this challenge in different ways with some success and failure. Many Corporate leaders have developed strategies and conducted a variety of analyses to gauge where their company is in comparison to:

- Their competition
- Assist in identifying and addressing operational and performance issues
- Identify new approaches to help move the company forward
- Retain a competitive position in their respective markets

Having knowledge of market position and how their business compares to competitors provides leaders with an opportunity to adjust their strategy and remain competitive and at times excel beyond the capability of their competitors. How to achieve the level of performance needed to sustain a business in the near and long term, and successfully compete in respective markets, is an ongoing challenge posed to many business leaders. The success of businesses depends significantly on the strategy that is developed and implemented, the leadership team, the culture of the organization, their ability to attract and retain talented employees, and the caliber of employees supporting the business. In order for a company to achieve their business objectives, operations must be strategic and sustainable. Included in a comprehensive strategy, there must be mechanisms to optimize and continue to improve performance across the board that include the human element, processes, and equipment. The desire to strive for improvement, coupled with a strategy that optimizes operational performance, is being referred to as Peak Performance for that company. Peak Performance can, and oftentimes differs, from company to company depending on the mission and goal of the company. It is even conceivable that Peak Performance can differ for companies with competing product lines in the same industry. What is Peak Performance and why does it matter? Peak Performance is achieved when an organization is able to remain highly competitive, recognized, and respected in the market in which they compete, delivers quality products and services to their customers and exceptional profits to stakeholders, and their employees want to work for the company because they are trusted and respected. A company that is functioning in the Peak Performance realm is capable of excelling in their respective business ventures, delivers exceptional value for their customers, and generates exceptional profits for stakeholders. Before an organization can reach Peak Performance, they must have the tools, processes, and practices in place that facilitate organizational effectiveness and stability. A successful organization will have processes in place to reduce and learn from human error, improve effectiveness and efficiencies, and continue to seek ways to improve.

It is recognized that Peak Performance is not easy to achieve; organizations must have a strategy and possess the staying power to impact and sustain implementation. Although not easy, it can be achieved and sustained using the tools outlined in this book. The right business strategy, coupled with a skilled leadership team and engaged employees, can help position a company to move toward achieving and sustaining their Peak Performance.

1.2 Strategic management

It has been said that "without a vision, the people will perish." This is also true for organizations and companies. All successful companies and organizations have a vision that is reflective of strategic management decision-making. Strategic management is a process that is ingrained throughout the company and critical to the success of organizations. The process encompasses the decisions and actions that form the basis of the plans that assist the company in achieving their objectives. The strategy contains the vision for the organization and the steps needed to achieve that vision. Most strategic management plans contain key attributes that include:

- Leadership support and decision-making process
- Lean thinking leaders (leaders that are always seeking efficient and effective ways of getting things done)
- Considerations for the external environment and stakeholders
- Recognizing that strategic actions can have multifunctional impacts
- Strategic actions must target the present and the future
- The organization's long-term viability depends on strategic actions and activities
- Allocating the resources necessary for success
- A plan and pathway for continuous improvement
- A learning organization

1.3 Workplace dynamics

Every company must deal with and adjust to workplace dynamics. More challenging is how to integrate the diversity of thoughts and cultures into a consolidated unit. Dealing with these dynamics is not straightforward and, most of the time, is tricky. Addressing workplace dynamics is not easy because it involves dealing primarily with three broad categories of behavioral attributes that form the basis of what guides behaviors in an organization (Figure 1.1). These attributes can represent competing priorities and instructions, and can be implemented in a way that seems counter-intuitive. When this occurs, it adds complexity to the prevailing workplace dynamics.

Problem-solving can become an issue when the workplace dynamics are not aligned with the goals of the organization and the members of the organization are not engaged and do not have the appropriate relationship with each other and the leadership team. It can also become an issue if communications and trust are not fluid within the organization. Consider the following scenario; the leadership team of Company X is implementing a strategy that will streamline and increase efficiency in their procurement and receiving processes. The company conducted a proof of concept pilot-test to ensure that the process will perform as expected. The result of the pilot-test demonstrated that the concept will increase efficiency by 40% and require less human capital to implement. The leadership team communicated to the affected workers that no one will lose their job and some workers will be reassigned to other groups within the company. Employees were skeptical of what was communicated to them because management had not always followed through with decisions in the past. As a result, many of the workers began to seek employment elsewhere. This exodus of skilled workers presented additional challenges since new employees had to be sought, hired, on-boarded, and trained to perform

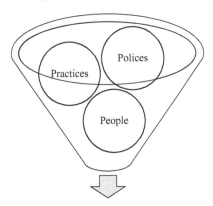

Figure 1.1 Workplace dynamics attributes.

critical tasks that were being performed by the employees that abruptly left the company. Workplace dynamics can facilitate good performance of a team to become extraordinary or cause them to fall apart. Improving workplace dynamics greatly depends on the ability to improve human relationships and behaviors. Workplace dynamics can negatively or positively impact an organization's ability to reach their Peak Performance.

1.4 Peak performance model

It is postulated that Peak Performance will be achieved and sustained by companies when a strategic process has been established and implemented. The Peak Performance Model is one such process that, if effectively utilized, can improve a company's performance and distinguish the company from the competition. In fact, it can drive operational performance from average to excellence. There are four functional elements of the Peak Performance Model:

- Lean
- Human Performance Improvement
- Operations Excellence
- Organization Culture

Individually, and collectively, these components will produce changes that can lead to incremental improvement in organizations. Often the changes or improvements realized are not sustained and seem like the flavor of the month to employees. However, used in unison these elements provide an organization with the best recipe for success toward achieving and sustaining their Peak Performance.

1.4.1 Lean

Lean has been tested and proven to be an exceptional process in improving performance by many companies. For example, Lean has been used successfully in the health-care industry, government, research and development, engineering, and more. The process and tools associated with implementing Lean are not optimally effective if not supported by a Lean-thinking leadership team. A Lean-thinking leader is strategic in thoughts and actions and supports a strategy of continuous improvement that is aligned with, and integrated into, the way business processes are defined and work is accomplished. There are some distinct behaviors one would expect to see from Lean-thinking leaders. These behaviors are depicted in Figure 1.2[1]. The behaviors exhibited by leaders are important attributes that have the ability to shape the way employees act, react to work situations, perform work, and their overall commitment to the company and

Chapter one: Defining peak performance 5

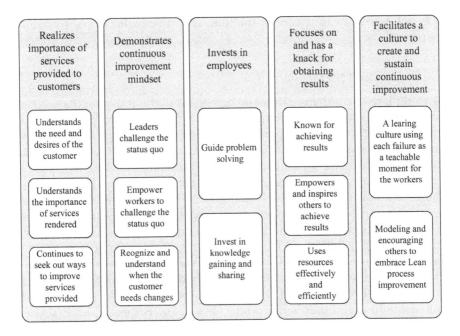

Figure 1.2 Lean-thinking leadership behaviors.

organization. The behavior of leadership is also important in building organizational trust, which is a key component of building relationships between organization members. Leadership behavior that builds trust exhibit the following attributes:

- Fluid communications that are honestly communicated with care
- Leading by example
- Admit mistakes when they occur
- Self-reflective and uphold commitments
- Demonstrate trust for organizational members
- Treat employees fair and equally

There are many tools available in the comprehensive Lean tool-kit that can ensure optimal implementation of Lean in an organization. Some of the tools that should be part of the tool-box include root-cause analysis, just in time, value stream mapping, elimination of waste (travel time, inventory challenges, excess motion, waiting periods, over-processing, overproduction, reworks, and human capital), five-step process (identify, measure, analyze, improve, sustain), process-flow mapping, and Supplier input Process output Customer (SIPOC). Chapter 2 provides more details on Lean tools. The tools mentioned represent a comprehensive list, but there are other tools that are available for use if desired.

It is not necessary to use all of the tools in your box for every situation or process.

1.4.2 Human performance

A workplace or company cannot be successful without the people needed to accomplish work. The interactions and behaviors of employees are a critical part of workplace dynamics. Understanding the types of behaviors that can be present is the first step in identifying methods to address these behaviors. There are a number of positive human behaviors that play a role in the way work is accomplished; some behaviors may be viewed as more favorable to productivity than others. Most behaviors seen in the workplace can be organized into one of three categories (Figure 1.3). Some behaviors are more obvious than others and some behaviors are more destructive than others. On the other hand, some behaviors are more favorable at improving productivity, work quality, and overall organization life balance. Getting work accomplished in the workplace involves the actions of humans, and the way they perform the work tasks and interact with each other is important for success. For example, an assertive employee may be preferable when working in a team setting where each team member is expected to be responsible for functioning as a subject-matter expert and provide guidance to the team and the team lead on critical project elements. An assertive team member will most likely be actively engaged in listening to input from other team members and should be able to express themselves without offending others; whereas, an aggressive team member is less likely to be heard, less likely to respect the contribution of others, and most likely to dominate input and decisions while alienating other team members.

1.4.2.1 Human behavior and emotions

In society, humans understand that emotions guide behaviors and behaviors impact decisions and actions. As such, it is important that individuals are able to recognize when their emotions are leading them into areas that can negatively impact the way they will react to a given situation. This awareness can help in controlling emotions and behaviors or actions taken by an individual. The impact of human behavior and emotions cannot be underestimated as to the effect they have on employees and their job performance in the workplace.

We often hear the phrase "check your emotions at the door." For many employees, this is an unrealistic expectation. The next best thing is to understand when our emotions are changing and gain control of the subsequent action or reaction that follows. Emotions are used during thinking, problem-solving, relationship-building and -retention, and the ability to trust others. Therefore, emotions are always at the center of who we

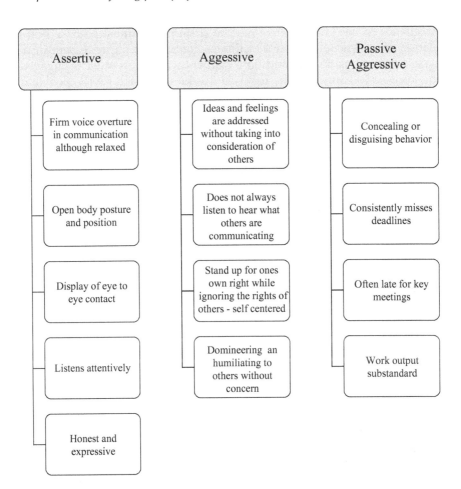

Figure 1.3 Characteristics of human behaviors.

are and the decisions we make at home and in the workplace. Behaviors fueled by feelings can impact everything, to include an individual's ability to make effective and constructive decisions, how creative they can be in problem solving, as well as their ability to follow instructions and directions from superiors as well as colleagues.

1.4.3 Operations excellence

Defining Operations Excellence in general terms is not easy because the definition varies depending upon the type of business or organizational structure. Some define Operations Excellence in terms such as the best in their markets or sector, the best in the world, world-class in the services and products produced, and the list goes on. In order for Operations

Excellence to become a reality, and be sustained, it must be based on a platform that encourages and facilitates continuous improvement in all aspects of the business. In addition, workers at all levels must be engaged; the leadership team must be forward-thinking, trusted by the people in the organization, and viewed as respected role models that others are willing to follow.

Critical components of Operations Excellence that enable Peak Performance include culture, ability to attract and retain a qualified workforce, defined programs and processes reflective of rigor and discipline in performance, innovative thinking, continuous improvement, a Lean-thinking leadership team, and employee engagement. Together, these components are key in positioning a company to reach the realm of performance that will yield the best output. Essential contributions of each of these components are listed in Figure 1.4. Key components of or the primary tenants of achieving excellence in operations in broad categories involve the caliber of the leadership team, the types of process and procedures in place, and their continuous improvement strategy that is supportive of streamlining operations. When seeking to achieve excellence, keep the following in mind:

- Design continuous improvement in the process
- Create and document work-flow standards for consistency in operations
- Empower employees to participate in development of the business strategy

Operations Excellence is defined differently by many, and the various definitions can be based upon the type of business, the goals and objectives of the leadership team, and the vision set forth to be accomplished.

1.4.4 Organization culture

The area of organizational culture has been studied by many practitioners and researchers for decades. As a result of these studies, the importance of organizational culture has moved to the forefront. Organizational culture has become a common topic of discussion for corporate leaders because it is widely recognized that culture plays an integral role in the effectiveness of the work environment. The culture of the organization has an influence on how people behave and are treated as well as how they perform and integrate into the work environment.

An important aspect of being able to enter into the realm of Peak Performance is having the appropriate organizational culture that can serve as the foundation in which work is accomplished, decisions are made, and engagement of the entire organization is viewed as constructive

Chapter one: Defining peak performance

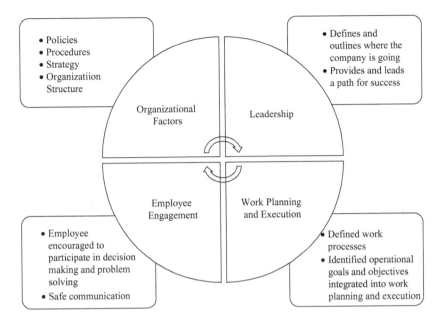

Figure 1.4 Operations Excellence components.

and necessary. The relationship between organizational culture and organizational performance is shown in Figure 1.5.

The culture of an organization facilitates behavior that promotes positive organizational performance. According to Alston[2], an organization's culture can substantially influence an organization, because of the shared values and belief within a culture that represent variables that guide behaviors. The culture of an organization also influences the behavior of people, their actions and reactions to various situations as they occur. An

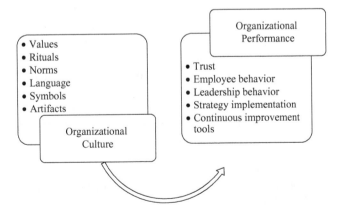

Figure 1.5 Organizational Culture and performance relationship.

extensive literature review was conducted by and documented in Alston, concluding the following key points:

- The culture of an organization can facilitate positive economic value for a company.
- A company's culture can be a viable source of sustainable competitive advantage if the culture itself is valuable.
- Culture has an impact on people's performance and therefore organizational performance.
- Culture determines behaviors of workers.

The performance of employees is tied to the cultural environment in which work is being performed. One can say that culture is the "bedrock" for organization performance and sustainability.

1.5 The impact of behaviors on organization performance

Employees are an integral and necessary part of efficiently running a business in order to achieve the goals of the organization. However, it is recognized that employers will make mistakes and mistakes can be costly to a business. Focusing on and understanding how, when, and why people make mistakes can provide an avenue for leaders to:

- Design their processes
- Develop programs and procedures
- Institute work practices that can minimize the potential for mistakes to occur
- Reduce the risk of having to deal with a catastrophic event

The behavior of employees can negatively or positively impact the performance of an organization. For example, in an organization where workers are inconsistent and display a lack of focus and concern for completing work safely, workers can become complacent, injured, or possibly die from a work-related injury. The money and time associated with workplace injuries go beyond just compensating the injured employee but also include injury investigation, event analysis, communications, negative publicity, and increased oversight by stakeholders. Development of solutions to prevent reoccurrence can also be costly in many ways. The company in some cases may have to cease operations and retrain employees or modify operational or business processes. In addition, consider the time that may be necessary to communicate to the public, stakeholders, and employees. The behaviors of workers and the leadership team are critical in guiding culture and work performance.

1.5.1 What do we mean when we refer to human behavior?

Human behavior encompasses the way a person presents themselves through their actions and interactions with others. These actions and reactions include all verbal and nonverbal communications. Our behaviors determine how we respond to family and friends, coworkers and management, react under pressure or stress, perform our work tasks, our ability to focus, and the way we respond to customers.

The key to modifying behavior is to understand the history and motives of people. This concept has been studied for decades by many. The answer to what motivates people is a question with answers that vary depending on the individual. One can say understanding the behavior of humans is complicated, and it is. Behaviors are generally learned by individuals resulting from some type of pattern or activities, or events that influenced their life. Therefore, it is feasible to modify a person's behavior by replacing unwanted behavior with a more desirable one. However, behavior modification is not always successful; or, if successful, it can be short-lived. Long-term new behaviors may require continual reinforcement for them to be sustained.

1.5.2 Leadership behavior and style

Leadership plays an important role in shaping an organization, and leaders have the ability to set the stage, and path, that will dictate the way the organization responds to all aspects of the business, including performance and success.

Leaders are responsible for charting the path of an organization as well as modeling the behavior that is expected of others. Some important contributions of leadership that can impact behavior include.

- Serving as a role model through respective behaviors
- Charting the path of the organization through strategy development and implementation
- Exhibiting trustworthy behaviors in actions, practices, and policies implemented
- Following all procedures and prescribed work practices

The behaviors exhibited by the leadership team can make or break an organization in terms of productivity and the organization's ability to achieve high performance and reach their Peak Performance.

The style of a leader will dictate the behaviors exhibited by the workers and the manner of communication exhibited by leadership permeates the organization, and others start to exhibit those behaviors. Leadership behavior and style can directly impact the performance of an organization in different ways. There are three predominate styles that leaders tend to exhibit and follow. These styles are: authoritative, democratic, and laissez-faire. An authoritative leader is commanding and dictates procedures and all of the decisions in an organization. As a result, the organizations are limited in the way they function, creativity is stifled, productivity can be hampered because of the lack of diversity of input, employees are not engaged in problem-solving and become uncommitted to the organization, and the list goes on. These leaders can be recognized through observation of the characteristics listed below:

- Little to no feedback sought from employees.
- These leaders tend to use the word "I" significantly in communicating.
- Decisions are made, directed, and enforced by the leader.
- Takes credit for successes.
- Lack of trust by employees.
- A disregard for the feeling and passion of others.

A democratic leader will encourage input and feedback from others to include employees, managers, and stakeholders on matters involving the business. This style of leadership, when present, is capable of gaining the cooperation of employees and moving the organization forward in a positive way. A democratic leader is able to actively engage employees and harness the best ideas and solutions to problems, has the ability to develop and implement strategic plans that can improve productivity and efficiency, and is able to gain trust and respect from employees. Some of the characteristics that may be seen in these leaders are:

- Open communications
- Generally trusted by workers
- Seeks to involve others in the decisions that are made on behalf of the organization
- Encourages participation and engagement of others
- Viewed as being approachable and open to changes

A laissez-faire manager takes a hands-off approach to leading and handling organization matters. When this type of management style is in charge, managers and workers are allowed to work according to their own beliefs, style, and vision. In organizations where this leadership style is prevalent, the following characteristics are generally observed:

- Low to poor motivation of workers.
- Manager is viewed as being absent and unengaged.
- Managers are not consulted when decisions are made.

- Workers are solving issues on their own with no input from the manager.
- Power is distributed to workers and the manager takes responsibility for their decisions and successes.

It is pertinent that the leadership style exhibited by leaders be known and training provided when leaders don't have the skills or characteristics needed to accomplish the goals of the organization. For example, in situations where workers are unfamiliar with a process or procedure, the leader may need to be intimately involved to coach and guide workers. In this situation, a laissez-faire manager style is not preferable to enable success in these situations. The use of a more democratic management style will yield more success since democratic leaders are more likely to coach and engage workers in problem-solving.

1.6 Organizational performance

The ability of an organization to achieve their objectives and excel is interconnected with the overall performance of the employees, leadership team, and the organizational strategy. Therefore, it makes sense that management focus on strategic ways to improve the performance of the workforce, and processes and systems that form the basis of accomplishing work. Generally, when looking at how well an organization is performing, a correlation is made between the desired output and the actual output. The desired output is supported by the vision, mission, and goals of the organization.

The primary means by which the performance of the organization can be evaluated is through a comprehensive assessment program. Assessing organizational performance is a critical aspect of a company's continuous improvement strategy and achieving Peak Performance. Managers must know how well their organizations are performing in order to determine what, and when, changes in strategy are warranted.

1.7 The peak performance model

The Peak Performance Model introduced in this book has important implications for organizations that have the desire to progress, improve performance, and sustain high performance over time. The model itself evaluates the value of three continuous improvement programs, along with the culture of the organization, to identify areas to improve and rise to Peak Performance. The model focuses on three continuous improvement programs: Lean, Human Performance Improvement, and Operations Excellence. In addition, the model recognizes the influence of organizational culture as a foundational element of a successful organization and company. By applying the Peak Performance Model, an organization or company will not only improve operational performance but will also exhibit the following characteristics:

- Lean-thinking leadership team.
- Management demonstrates concern for employee health, wellbeing, and safety.
- Employees are creative in problem-solving and do not accept "we have always done it this way."
- Trust permeates throughout the organization at all levels.
- High employee engagement.
- Employees have ownership in the overall success of the company because they participated in the development and implementation of the company's strategic plan.
- Employees feel free to communicate with management on issues as they arise.
- The culture of the organization is one that facilitates high performance and collaboration.

References

1. Alston, F. (2017). *Lean Implementation Application and Hidden Costs*, CRC Press, Taylor & Francis Group.
2. Alston, F. (2014). *Culture and Trust in Technology-Driven Organizations*. CRC Press, Taylor & Francis Group.

chapter two

Principles of lean to optimize business outcomes

2.1 Introduction

How is Lean typically defined? Lean is defined as a technique that employs a systematic method or process for identifying and minimizing waste without negatively impacting or sacrificing production and quality. Lean takes into account waste that is created through various activities, to include overburden and irregularity in workloads. The overall objective of Lean is to focus on the clients and consumers of a product or service by providing value that they can buy into. Value consists of any action or process that goes into developing the final product or services that a customer would be willing to pay for.

How did this process start, where did it start? Lean principles were derived from the Japanese manufacturing industry. The term was invented by John Krafcik in his 1988 article, "Triumph of the Lean Production System," based on his master's thesis at the Massachusetts Institute of Technology (MIT) Sloan School of Management.[1] Krafcik was a quality engineer with Toyota–GM joint venture in California before joining MIT for his MBA studies. His research continued through the International Motor Vehicle Program (IMVP) at MIT, which produced the international best-selling book co-authored by James P. Womack, Daniel Jones, and Daniel Roos (1990) called *The Machine That Changed the World*.[2] Lean is a philosophy and an approach that stresses the elimination of waste or non–value-added work through a focus on continuous improvement to streamline the operations. It is customer-focused and stresses the concept of eliminating any activity that fails to add value to the creation or delivery of a product or service. Lean is focused on providing higher quality, reduced cycle time, and lower costs. Because Lean results in improved performance in production and company systems, it is believed to increase production capability and flexibility. The birth of Lean started with Toyota's philosophy and the development of the Toyota Production System (TPS). Toyota desired to be the best in the United States Automobile Industry, a goal they obtained despite a lack of resources and infrastructure. They achieved this goal through the application of Lean Principles and many of the Lean Manufacturing Tools. Is Toyota the only industry that has achieved success through the use of Lean Principles? No. Many companies have successfully implemented a robust Lean process – but have these companies reached Peak Performance? Let's explore the Lean process and tools to obtain a better understanding of the overall process and potential benefits.

2.2 Why lean?

The real purpose of Lean is to eliminate waste in the process and maximize resources, by looking at the process and asking a simple question: What does the customer want/desire? If it is not adding value to the customer, then it is considered to be waste. If the customer does not explicitly want it and it does not render the product or service better, then ask why are you doing it. For example: if you package your product in both plastic and a box, and the customer provided feedback that they only need one of the packagings, then the plastic or box may be considered waste, and value is not being added. The other driver for Lean is to become as cost-effective as possible; this is a benefit to both you and your customers.

It is important in any process to understand "why" things are done in the manner in which they are performed. Using the example discussed earlier – packaging products in both plastic and a box – you should ask yourself why. Maybe you have always done it that way or maybe the product loses its freshness if only packaged in a box, or the product can easily be damaged without a box. These are important pieces of information to consider when moving forward.

One of the most critical elements of implementing a Lean program is the culture of the organization. If your organization does not have a learning culture, the ability to adapt to change, and a culture where workers are respected and engaged, then implementing a Lean process will be very challenging.

Lean is best implemented when small but effective changes are made, giving everyone a chance to see and understand how little changes can make a big difference. Once implemented, with the goal of continuous improvement, other areas should be identified for change, and the improvement process will continue to grow, with others engaged in identifying areas of waste. The thought process will start to be ingrained in their minds. The leadership team is the key to setting the tone and showing ownership of Lean. Making time to listen and watch what is really happening in the facility or office is a good start. This allows the Leader to see first-hand what is taking place, what issues the team is having, and, just as important, it sends the team the message that the leaders are taking the time to identify and observe challenges, listening to and understanding their needs, taking an interest in jobs they perform every day. It takes time for employees to embrace a continuous improvement process and environment, but with time, practice, and a good leadership team, it can and will happen.

The implementation of a Lean program or process is through preventing and eliminating waste in processes. Waste is categorized as travel time, inventory challenges, excess motion, waiting periods, over-processing, overproduction, reworks, and human capital. All of these wastes have a direct impact on productivity and costs; these are all non-value-added activities, operations that your customer does not want to pay for and adds no value to the product or service provided. It is recognized that waste can be found in most processes, but what if we could eliminate 95% of wasted time and effort. What would that do for the organization, how much money could be saved? The savings could be reinvested in production, reduce the cost of products, and significantly improve the bottom line! Once you start to recognize the forms of waste, you will start looking at your facility and processes from a different viewpoint, with what we call a "Lean Lens." You will begin to see your processes and inventory in a whole new light, one that you will be eager to start changing. But remember to focus on incremental, daily changes that can result in significant improvements over time. Start where you believe they will have the most impact and where employees are the most engaged.

2.3 Evaluating the waste in the process

Let's evaluate each form of waste and how it can be applied to a business or production facility. Learning these tools will provide the insight needed to start assessing a business or an operation to eliminate waste. Now when you walk through your business and/or facility, you will have a whole new perspective on each operation, each facet, asking yourself: Is this part of the business adding value, or just creating more waste? When you look through a "Lean Lens," you will start developing a questioning attitude about all parts of the business, continuing to ask "why" until you receive a solid response that makes business sense. This will start you down the road to improving the bottom line and contributing to a learning organization and culture.

2.3.1 Travel time

Travel time is the movement of materials from one location to another, which is considered waste, as it adds zero value to the product. The waste of travel time can reduce productivity and be a very high cost to an organization. Travel time adds no value to the product; people are compensated to move material from one location to another, a process that costs money, adds a significant opportunity for injuries, and provides no value.

In an office setting, multiple review points or hand-offs, or multiple people copied on an email instead of the ones who actually need the information to complete a task, can be considered waste. Employees having to travel to get copies, supplies, etc. can reduce the productive time available and efficiency. Figure 2.1 depicts travel time as waste in an operational setting. Why would your customer want to pay for an operation that adds no value? There are several ways to reduce travel time: for example, store the materials as close as possible to the operation using it. Utilize just-in-time shipping of materials, again eliminating the extra storage needed and extra travel time required to move the product. Have the vendor or operations personnel place the materials where you need them to be stored. Now that you have reduced the travel time, you can save money by reducing the storage space required, and reduce the workforce needed to move the materials. This may also improve worker safety.

The Bureau of Labor Statistics reported in 2016 that 18% of the injuries requiring days away from work involved material handling. If you have an employee injured and unable to come to work, it's costing the company not only in worker compensation but also in paying another employee to cover the workload until the injured employee can return to work. How much could an injury impact your bottom line?

Figure 2.1 Example of waste from travel time. (Courtesy of PRESENTERMEDIA.)

2.3.2 Inventory

Inventory costs money; every piece of material tied up, lying around, work in progress, or finished product has a cost until it is sold. In addition, to the pure cost of inventory it adds multiple other costs, feeds many other wastes streams. Figure 2.2 depicts inventory that has to be stored and needs physical space, storage racks, carts, or bins, someone to maintain the inventory, and transportation to move it around the facility. Some inventory may need special care, for example, temperature control to prevent freezing, humidity control, or cleanliness control. Inventories need managing to prevent them becoming obsolete, to ensure packaging remains intact, or to avoid them being damaged during movement/travel.

The quality of inventory can deteriorate over a period of time, especially perishable items such as food or rubber seals. The waste of inventory challenges may mask other wastes in the systems.

The next time you are in your facility conducting a walk-down, look at the materials with a "Lean Lens." Do you see excess materials in the workspace, are employees hoarding supplies, do you see frequent movement of materials, are the materials in date, or is there any damaged inventory? Look at the waste dumpsters. What are you seeing in them? Good inventory that is being discarded for some reason? Maybe there was no room in the storage area. You can learn a lot about the facility by looking in the waste dumpsters. Evaluate the requirements for material storage. Are materials stored according to manufactures recommendations? Estimate the overall cost of your inventory – that's *all* considered waste. Your customer does not want to pay for a company's inventory challenges, nor does your company.

2.3.3 Excess motion

Excess motion is the movement of people or equipment, made while working, that may not add value in creating products or services. As depicted

Figure 2.2 Excess inventory storage. (Courtesy of PRESENTERMEDIA.)

in Figure 2.3, this includes walking, lifting, reaching, bending, stretching, and repetitive movements. For example, excess motion could be moving a compressor unit from a top shelf in a maintenance shop instead of storing it on a lower level to reduce stress to employees and time to access the equipment, like having to walk to another area to obtain a part, tool, or document such as a procedure. Another example is sorting through an unorganized area to look for a document or parts. The movement can also involve employees walking between areas to talk with colleagues, walking to a printer, photocopiers, filing cabinet, storage cabinets, readjusting a component after it has been installed, sifting through inventory to find what is needed, excess mouse clicks, and double entry of data. Excessive travel between locations and excessive machine movements are also examples of excess motion. All of these wasteful motions cost money and time and can cause injuries to employees and stress to equipment and processes

The next time you are conducting a facility walk-around, look at the movement of products and resources with a "Lean Lens." Observe the following: are the workspaces well organized; is the placement of equipment

Figure 2.3 Excess Motion – storage of heavy material. (Courtesy of PRESENTERMEDIA.)

Chapter two: Optimizing business outcomes 21

near the production location, storage of materials and supplies (heaviest materials and equipment should be stored closest to the floor level – see Figure 2.3); and are materials stored in an ergonomic position to reduce stretching and straining?

2.3.4 Waiting periods

Waiting periods can be the most obvious aspect of waste; they are easily identifiable as lost time, potentially due to poor set-up such as a part shortage, bottlenecks in the process, having equipment with insufficient capacity, waiting on a cycle to complete, and equipment failures. It is the time spent waiting for a response from another organization, waiting for software to load, waiting on email, waiting on a signature, ineffective meetings, waiting for a procedure revision, or waiting on repair of a piece of equipment. Waiting periods are often caused by an imbalance in the production areas that can result in excess inventory and overproduction. Waiting periods can cause frustration among workers, which can lead to reduced morale, lower productivity, and workplace injuries. The goal should be to design processes to operate automatically where feasible, and to ensure that standard work practices have the worker arriving just after the machine cycle has completed, and not before, to avoid wasted waiting periods.

When conducting a walk down, look at movement with a "Lean Lens." Are processes designed to ensure continuous production/flow; are workloads optimized by using standardized work instructions; are workers arriving at a work station just after the equipment has completed the cycle; and are workers flexible and multi-skilled to quickly adjust to work demands?

2.3.5 Overproduction

Overproduction is perhaps one of the costliest of all wastes; the waste is generated by making too much product and/or making it too early. Overproduction occurs when producing a product or a segment of the product before it's required or requested. It is producing more product than the customer ordered and will pay for. It may be tempting to produce as many products as possible when there is idle time or worker availability. The excess product has to be stored somewhere, which means excess motion, travel, and inventory challenges. Remembering that waste is anything for which the customer is not willing to provide compensation, it is easy to see why overproduction is a costly waste. Also, overproduction means that if a reject is found, there will be more units that need to be reworked. However, overproduction can drive all of the other types of waste as well.

In an office environment, for example, overproduction could include making extra copies, creating reports that no one reads, providing more information than needed, and providing a service before the customer is ready. Additionally, overproducing a product also leads to an increase in likelihood that the product or quantities of products produced are beyond the customer's requirements.

Again, look at the company's waste storage areas: are there discarded products made at the beginning or end of a product run, maybe even a good product that was discarded because it did not fit on a pallet or in a storage area? When conducting facility walk-downs, ensure that you look at the overproduction with a "Lean Lens" and explore the following questions: are processes designed to ensure continuous production flow and the rate between work stations is uniform; do you see excess product stored in the shop or in dumpsters; are workers drafting reports that never get read? Looking through a "Lean Lens" can be informative and provides information that can be used to facilitate process improvement activities that will reduce or eliminate the waste generated.

2.3.6 Over-processing

Over-processing refers to doing more work, adding more complexity or components, or having more stages in a product or service than what is required by the customer. In production, this could include using higher accuracy than necessary, using modules with capacities beyond what is required, running more analysis than needed, over-engineering a solution, adjusting factors after installation, and having more functions than the product needs. An example is fabricating a product that will last for five years when the customer is going to replace it after three years. Figure 2.4 presents a visual impact that may be seen on the shop floor when overproduction occurs. What is not visible is the wasted resources used, along with the funding expenditures that could have been used to improve operations in other areas of the process.

Remember to continue to use a "Lean Lens" when walking through the facility and observing work being performed, and that the customer only wants to pay for what is required!

2.3.7 Rework

Rework occurs when the product or service is defective or does not fit specifications required by the customer. Consider as an example, car parts that do not comply with the specifications of the customer. If the part is not correctly fabricated it will require more time to correct, either by sending the part back to the assembly line or, at worst, having to completely scrap the part and begin from scratch. What if incorrect data requires rework

Chapter two: Optimizing business outcomes 23

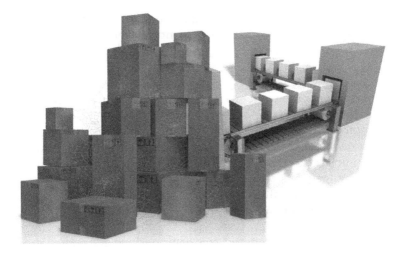

Figure 2.4 Overproduction. (Courtesy of PRESENTERMEDIA.)

of engineering calculations and rework of custom component? These are wasteful, and add additional costs to the operations without delivering any value to the customer. It is much better and more efficient to make the components or provide the service right the first time, which builds a better relationship with your customers as well as with the workforce.

The next time you are in your facility conducting a walk-down, look at rework with a "Lean Lens," look for the most frequently reworked component, that part of an assembly line that always seems to have a problem or service calls that must be repeated, then focus your attention on that area. Try to determine if the cause of the defect is human error, equipment failure, or a combination of the two.

2.3.8 Human capital

Human capital is considered by many a form of waste in a process. One can describe this form of waste as underutilized human talent and innovation, or ineffective use of time. This can occur for many reasons, such as stove-piped functions or organizations, or unclear roles and responsibilities. In some organizations, the leadership's responsibility is planning, organizing, controlling, and innovating the production process. The employee's role is to follow orders and execute the work as planned. This mindset creates an environment for waste to flourish as a result of not fully taking advantage of the diversity of knowledge and thoughts available to the organization.

In a production facility, this waste can be seen when workers are inadequately trained, not knowing how to effectively operate equipment;

given the wrong tool for the job and not recognizing it; when workers are not challenged to come up with ideas to improve the process; or when the workers have unproductive time. By not engaging the worker's knowledge and expertise, it is difficult to improve processes, because most leaders aren't as knowledgeable of the process as the workers that operate it. The workers doing the work are the ones who are most capable of identifying problems and developing solutions. Waste in an office setting could include: underutilized talent resulting from inadequate training; lack of worker motivation; not asking for worker feedback; idle time; and placing workers in positions below their skills and qualifications levels.

Worker engagement is an integral part of making Lean a successful part of a business. A Gallup poll of US workers in 2015 showed that just 32% of employees were engaged. A majority of employees (50.8%) were "not engaged," while 17.2% were "actively disengaged." Having more engaged workers leads to better company performance; it means more efficient processes, lower employee turnover, and higher rates of innovation and creativity. Engaged employees believe they have a direct impact on the company's performance and are more likely to try and invest in new ideas. Additionally, organizations with more engaged employees can achieve higher competitiveness, enhance customer satisfaction, and embrace a learning culture of solving problems through teamwork.

The next time you are in your facility conducting a walk-down, look at the use of human capital with a "Lean Lens": are your employees engaged; do you see them actively working or wasting time; what types of recommendations are you receiving from your employees and how are you providing feedback; do you see your leadership team asking for recommendations to improve and innovate the process? Ask your employees if they believe they are being utilized to the fullest extent: are they really working up to their potential; do you have a program to encourage workers to obtain college degrees or certifications, etc.?

2.4 Lean tools

Now that you understand the benefits of Lean and the various types of waste that are associated with business, let's explore a five-step action process, associated with Lean, that can be used to make changes in your processes. Remember you will need all employees involved in Lean activities, from the CEO to the shop-floor worker, to make suggestions and seek ways to improve the tasks they perform on a daily basis, thereby improving the business as a whole. When a Lean process is correctly implemented, the manner in which work is performed is improved and a creative work environment is fostered, one where employees are willing to become engaged, feel empowered, and recognize the value of their

Chapter two: Optimizing business outcomes 25

contribution to the success of the company. A Lean process can help eliminate excessive work, ensure that the workplace is ergonomically designed to avoid injury to workers, and facilitate appropriate storage practices. Lean provides employees with the tools needed to spot and eliminate waste in business processes from the office to the shop floor. It provides a mechanism for the entire workforce to be engaged and helps them to feel like they have a "say" in what they do and a sense of ownership in the success of the company. When Lean is implemented, it will be surprising how many great ideas will be presented to management from the workforce that will greatly improve business operations.

2.4.1 *The five action-steps to lean*

During one of your facility walk-downs, using your "Lean Lens," seek to identify areas that are generating waste. The following five-step process can help improve the identification process to sustain the measures put in place to transform practices. Each of these steps shown in Figure 2.5 will be discussed in detail in Sections 2.4.1.1 to 2.4.1.5. The five-step process presented in this book is a modification of the traditional Lean five-step process.

2.4.1.1 *Step 1 – Identify the problem*

The first step is to identify and clearly define the problem associated with a work process you need to resolve. Remember to involve the workers when you are trying to identify the problem because they are typically more knowledgeable about implementation of work processes and have

Figure 2.5 Five action-steps to Lean. (Courtesy of PRESENTERMEDIA.)

good ideas on how to improve operations. Form a team of all impacted groups and ensure that you have a representative from each of the stakeholders to enable a comprehensive solution to be developed. Select a team lead, one that is going to be willing to listen to all inputs and develop a solution that everyone can live with and that meets the customer requirements and needs. After the team is formed, begin by generating a map of what the work process looks like currently, i.e., how things are really done. This is probably one of the most critical steps in the five-step Lean process, so don't rush it, it needs to include all inputs and be the "way things are done" currently. Let's look at an example of a map for ordering a new wrench (see Figure 2.6).

The team make-up should include all stakeholders, such as a worker that could request new equipment, engineering, supply chain, receiving, and Quality Control (QC), participating on the team to ensure accuracy and completeness. There are multiple tools available to identify the problem and help develop a process map, including value-stream mapping, A3, and Supplier Input Process Output Customer (SIPOC) templates.

Value-stream mapping, also known as VSM, is a Lean tool for studying the current state and creating a future state for the sequence of events that take a product or service from its beginning through delivery to the customer. Value-stream mapping is an excellent tool to flush out waste in any process, not just production. Each major step of the targeted work process is itemized and gauged on its increasing value – or not – from the customer's standpoint. VSM also provides an excellent tool for communications, partnering, and potentially even facilitating culture changes. The emphasis on value keeps the analysis focused on what really matters to the organization and keeping them competitive in their market.

Anticipating or confronting competitive pressures, Lean practitioners utilize VSM to generate the most value to their customers in the best and most efficient manner. Once a process is mapped out, things become clear; for example, likely process delays, constraints, excessive downtime, and inventory issues, to name a few. VSM should be utilized on an ongoing basis to forge ahead on continuous improvement, always yielding better solutions to explore. VSM allows you to see not only the waste but the source or cause of the waste. By developing the future state and/or ideal state it makes it easy to develop effective solutions to improve the process.

What is an A3 process? It is a systematic approach to problem-solving and continuous improvement by completing a simple and rigorous method. The A3 approach is divided into a number of steps to enhance problem-solving. Figure 2.7 provides an example of an A3 template to utilize to facilitate discussion and find solutions to problems.

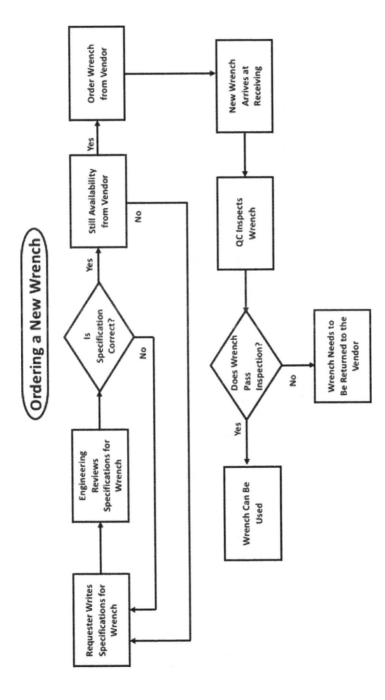

Figure 2.6 Process-flow map for ordering a new wrench. (Courtesy of PRESENTERMEDIA.)

Steps include:

1. Why are we here, the reasons we need to take action?
2. The initial state of the process
3. Goal – Where do we want to be?
4. Complete a gap analysis
5. What are the possible solutions?
6. Select the most viable options
7. Implement the changes
8. Confirm the changes are effective
9. Share what was learned

SIPOC, a process-improvement tool, is an acronym for Supplier Input Process Output Customer specifications.[3] The SIPOC Diagram is an easy-to-follow table that drives a team to focus on the correct process (see Figure 2.8 for an example). The SIPOC will also support the team's understanding of who the customers are and what are their specifications. It also helps divide the process into input and corresponding supplies, process steps, the outputs, and customers. The goal is to describe the process at a high level. The SIPOC is often used at the beginning of process-improvement efforts to support identifying or defining the issue. It has three standard uses depending on the audience. These uses are:

1. Provides an overview for workers that may not be acquainted with the entire process
2. Allows workers to review the process to jog their memories or to update new changes
3. Provides a platform to envision a change or new process

A3 Template – 9 Step Process		
1. Why are we here?	4. Gap Analysis	7. Implement
Why is it crucial to improve the process?	What is our gap between where we are and where we want to be?	Implement the changes identified in Step 6.
2. Initial State	5. Solutions	8. Confirm
What do we do well now?	What can we do to improve, what are the possible options?	Measure the change to validate the improvement.
3. Goal	6. Selection the Solution(s)	9. Share Successes
Where do we want to be?	Identify the solution(s) that is/are the most viable to drive improvement.	Share the lessons learned and the successes the team achieved.

Figure 2.7 Example of an A3 template.

Chapter two: Optimizing business outcomes 29

The SIPOC tool can be utilized when the following are not well defined:

- Who is providing inputs to the process?
- What requirements are identified for the inputs?
- Who are the actual customers of the process?
- What are the real specifications that the customers' needs?

2.4.1.2 Step 2 – Measure the targeted work process

Now that the process has been defined, it is time to start quantifying the issue, problem, or challenge. What is needed next is to really gain an understanding of how the process performs. In order to do this, data must be collected to validate the assumption or perception personnel have for how well or not so well the process actually performs. Being able to measure performance through the life of a project or facility is crucial. The data-collection process must include collection of data that shows how the process performs, as well as what the customer is really paying for. The team should focus on gathering data that will help reduce the overall time to make the product or deliver a service as well as how to maintain or improve the quality of the product or service. By measuring the current performance, the team can better recognize where changes or improvements need to be made.

To get started it's important to develop a data-collection plan: for example, what is going to be measured; how is the data going to be gathered;

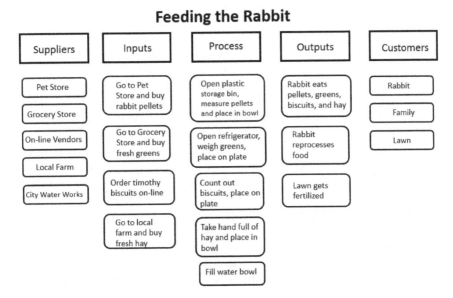

Figure 2.8 Example of a SIPOC.

how long the data will be collected; and how are you going to ensure the reliability of the data collected? The plan should include a short description of the project; what understanding the data might provide, and how it can support enhancements to the process; the justification for collecting the data; and what specifically will be done with the data. The plan should also capture the resources and activities necessary to accomplish data collection, and a detailed list of the specific measures needed.

Data can be obtained through a variety of methods including surveys, interviews, focus groups, observations, and document or records reviews. Let's look at a brief description of each data-collection method including advantage and challenges.

Surveys contain a consistent written set of questions that can be delivered by mail, email, or in person. The cost of conducting a survey is low and the amount of data collected can be high; it can also be conducted by multiple individuals, and the data obtained is typically easy to compare and analyze. There are some disadvantages to using a survey for data collection. Typically, if the survey is conducted using mail or email, the response rate can be very low, risking the validity of the data collected. Written surveys, which may be viewed as impersonal, don't allow respondents to obtain clarification on a confusing question that can bias the response depending on the wording. Survey examples already exist that can be modified to support data collection, but this can also take time. Tips that can be used to strengthen the use of surveys include conducting comprehensive testing of the surveys to reduce the probability of issues associated with the questions due to ambiguity that may introduce potential bias, and soliciting feedback from survey participants as soon as the survey has been completed.

Interviews are similar to surveys and should be conducted using a standard set of questions to obtain responses. Interviews can be conducted either in person, via email, or on a phone call. Interviews can provide the opportunity to be more flexible when it comes to having the ability to ask more complex questions because there is an opportunity to clarify confusing questions. This method also provides an opportunity to ask follow-up questions and to obtain more in-depth responses. Conducting interviews can be expensive and time-consuming; less data may be obtained because of the time it takes; data reliability can be a challenge if the interviewers are not well trained; interviewers can introduce a bias depending on how the questions are asked; and the data can be more difficult to compare and analyze.

Focus groups are small groups of people brought together to provide feedback on a defined topic. Focus groups are intended to allow participants to discuss the questions and share their experiences and opinions. This tool provides the opportunity for participants to influence each other, prompt memory, as well as provide an opportunity for participants to debate an issue. Focus groups provide a quick and reliable method to gain a response to questions, allow for in-depth information on a topic, and provide the facilitator with a sense of how well a topic is received by the group. The disadvantage of this method includes: the challenge of scheduling a time for individuals to meet; finding a good facilitator to keep the group on topic to produce the needed results; and the data can be difficult to compare and analyze.

Observations can be used to denote behaviors, observe interactions or events, and note site conditions. This method requires well-trained individuals that can follow a well-designed guideline for what or whom to observe, how long the observation should last, and how the data will be recorded. This tool provides validation of the data because it is obtained first hand though viewing behaviors and actions depicting how things actually occur. This method is expensive because of the time required to develop the tool and collect the data, as well as the trained, experienced personnel needed to conduct the observations. This method can also be challenging when comparing and analyzing the data.

Document or records-review involves methodical data collection from existing sources. The tool can provide an easy method to collect existing historical data with limited effort and minimal impact to stakeholders. This can be an inexpensive and efficient method to obtain needed data. The disadvantages of this method are: ensuring a clear understanding upfront of what data is necessary; the time it takes to review the information; the data is not flexible; and the data could be incomplete, leaving gaps making the analysis difficult. Table 2.1 summarizes each method of data collection, including the advantages and disadvantage.

2.4.1.3 Step 3 – Analyze the data

With the data collection complete it's time to begin the analysis to understand the causes of the issues, i.e., why the waste is being generated in the process. If this step of the process is truncated, it can prevent the root cause from being identified and the team to develop and implement solutions, therefore not really solving the problem or issue. During this step, the team should brainstorm ideas of the potential root causes, develop a theory as to why the problem or issue exists, and then work to prove or disprove their theories. There are several tools that can be utilized for this step including a box plot, fishbone diagram, and 5-Why.

The box plot is a method for describing groups of numerical data through quartiles, as shown in Figure 2.9. A box plot is a way to show the spread and centers of a data set.

Another effective tool is a fishbone diagram, also known as a cause-and-effect diagram. Fishbone is a visualization tool for classifying the potential causes of a problem in order to distinguish its root causes. A fishbone diagram (see Figure 2.10) is typically developed by an individual

Table 2.1 Comparison of Data-Collection Methods

Method	Advantage	Disadvantages
Surveys	• The cost of conducting a survey is low • The amount of data collected can be high • It can also be conducted by multiple individuals • The data obtained is typically easy to compare and analyze	• The response rate can be very low • May be viewed as impersonal • Don't allow respondents to obtain clarification on a confusing question • The response can be biased depending on the wording of the questions • Revising of existing survey questions can take time
Interviews	• Provide the opportunity to be more flexible in exploring responses • The ability to ask more complex questions • Chance to clarify confusing questions • Ask follow-up questions to obtain more in-depth responses	• Expensive and time-consuming to conduct • Provide less data because of the time it takes to conduct interviews • Data reliability can be a challenge if the interviewers are not well trained • Interviewers can introduce biases depending on how the questions are asked • Data can be more difficult to compare and analyze
Focus Groups	• Provides the opportunity for participants to influence each other, prompt memory, as well as motivate and clarify an issue • Provides a quick and reliable response to questions • Allows for an in-depth collection of information • Provides a sense of how well a topic is received by the group	• Challenge of scheduling a time for individuals to meet • Finding a good facilitator to keep the group on topic • Data can be difficult to compare and analyze

(*Continued*)

Chapter two: Optimizing business outcomes 33

Table 2.1 (Continued) Comparison of Data-Collection Methods

Method	Advantage	Disadvantages
Observations	• Provides validation of the data • A real view of how things actually occur	• Expensive because of the time required to develop the tool, have an individual spend time observing tasks, and collect the data • Finding a trained and experienced person to conduct the observations • Challenge to compare and analyze the data
Document and Record Reviews	• Provides an easy method to collect existing and historical data with limited effort • Minimal impact on stakeholders • Inexpensive and efficient method to obtain needed data	• Establishing a clear understanding upfront of what data is necessary • Time it takes to review the information • Data could be incomplete

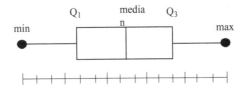

Figure 2.9 Example of box plot. (Courtesy of PRESENTERMEDIA.)

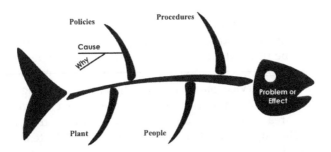

Figure 2.10 Example of a fishbone diagram. (Courtesy of PRESENTERMEDIA.)

that has knowledge of the process and systems. This tool is best suited for problems in which hard data is unavailable, or as preliminary work to identify potential causes that might be worthwhile for data collection and future analysis. There are four main steps to developing a fishbone diagram:

1. Identify the problem
2. Identify the major factors involved
3. Identify the possible causes
4. Analyze the diagram

The final tool being introduced is commonly used is referred to as 5-Whys. This tool is one of the easiest to use that does not require any statistical analysis. The method involves a practice that identifies a problem and asks the question why five times. By repeating the question why, one can peel away the layers of symptoms that can reveal the root cause of an issue. This technique is called the 5-Whys, but it may require asking the question "why" more or less than five times depending on the issue. This technique works best when the issue involves human interaction and/or human factors.

2.4.1.4   Step 4 – Improve the targeted work process

The root cause(s) of the problem or issue has been identified and now it's time to implement and verify the solution(s). During the improvement step, brainstorming solutions takes place, as well as implementing solutions and collecting data to confirm the improvement. Piloting changes on a small scale may be appropriate prior to wide-scale implementation. A structured improvement endeavor can lead to creative and innovated solutions that yield real changes in the overall process. Again, as for all of the steps, recognize that there are multiple tools that can be utilized for the improvement step, such as plan-do-check-act, swim-lane diagram, and weighted decision matrix to name a few.

2.4.1.4.1   Plan-do-check-act   As depicted in Figure 2.11, Plan-Do-Check-Act (PDCA) is a four-step model for inspiring change. The cycle should be repeated to yield continuous improvement in the process.

PDCA can be utilized when starting a new improvement project, for a repetitive work process, or when developing a new or improved process design, product, or service. Start by planning a process change, recognizing an opportunity to make a change; test the change; maybe pilot the change in one area of the plant, analyzing the results; and then ask if the change made the improvement you were seeking. Did the team learn anything that could make the process better? The last step is to take action

Chapter two: Optimizing business outcomes   35

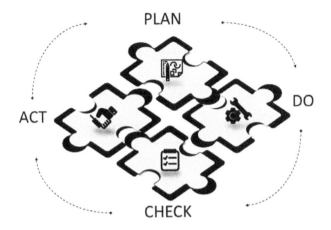

Figure 2.11 Plan-Do-Check-Act four-step model. (Courtesy of PRESENTERMEDIA.)

based on what was learned from the evaluation. Now use the cycle again and again to continuously improve and optimize the overall process.

2.4.1.4.2 Swim-lane   A swim-lane diagram is a type of flowchart, also known as a cross-functional flowchart, that depicts a process from start to finish, and divides the steps into groupings to help distinguish which team, group, or workers are responsible for each set of actions. The lanes are columns that keep actions separated visually from each other. A swim-lane helps make responsibilities more predominant than a typical flowchart. As depicted in Figure 2.12, a swim-lane diagram allows one to see not only the flow of the process but also the locations when responsibilities are transferred from one worker to another. Knowing which organization or worker is responsible for a particular step can help speed up implementation of corrective actions and eliminate delay and bottlenecks in the process. This method is also useful in helping organizations work cohesively together because each will understand what the other organization does to contribute to the overall process. This method can also identify redundancies and duplications, gaps, and communications issues between organizations or teams.

2.4.1.4.3 Weighted decision matrix   A weighted decision matrix is a tool utilized to evaluate process alternatives, weighted by importance to the process, based on specific evaluation criteria. Figure 2.13 presents the outline of a typical weighted decision matrix. By evaluating the alternatives based on their functionality with respect to individual criteria, a score (value) for the alternative can be determined. Assigned weighting factors

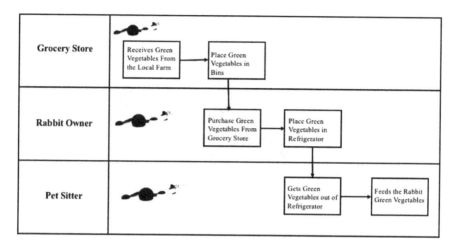

Figure 2.12 Example of swim-lane diagram. (Courtesy of PRESENTERMEDIA.)

	Criterion 1 Cost Effective		Criterion 2 Worker Satisfaction		Criterion 3 Flexibility		Total Score
Alternatives	Raw Score	Weighted Score	Raw Score	Weighted Score	Raw Score	Weighted Score	
Option 1							
Option 2							
Option 3							
Option 4							

Figure 2.13 Example of a weighted decision matrix. (Courtesy of PRESENTERMEDIA.)

for each criterion are based on the importance of each. For example, if cost is an important factor in evaluating alternatives, make sure it is assigned the higher weighting factor relative to each other criteria. Weighting factors could be, for example, a score of five (5) for a mandatory criterion, two (2) for a desirable criterion, and zero (0) for an optional criterion. To calculate the weighted score, multiple the raw score by the weighted factor and obtain the weighted score (Figure 2.13). The scores for each alternative can then be compared to create a ranking order of their functionality related to the criteria as a whole. This tool allows each criterion to be reviewed independently, helping to avoid the emphasis or over-influence on one specific criteria. Using the weighted decision matrix, an organization can help narrow options using specific criteria important to the project.

2.4.1.5 Step 5 – Sustain process improvement

Once the problems have been identified and measured to understand their magnitude, the issue has been analyzed, solutions have been developed and tested, and corrective actions implemented, it is time to ensure the process is able to sustain the gains from the implemented improvements. This last step of the five-step action process is focused on creating a monitoring plan to continue measuring the success of the improvements and developing a response plan in case there is a degradation in performance. As with all of the steps, there are several tools that can be used to monitor performance; examples include monitoring and response plans, documentation, and control charts. A monitoring plan helps uncover changes and when these changes may have occurred in the process, the monitoring plan confirms that improvements continue to be maintained, which enables requirements to be met over a long period of time. A monitoring plan can include the following:

- Key output process measures and gauges for continued improvement
- When data should be collected and how frequently
- Describes the method for collecting, recording, and reporting of data

A response plan lays out what actions will be taken if a deviation is detected during monitoring. The plan will define the timeframe when an action should be taken once a deviation is noted, who will take the action, and the proper procedure to use to rectify the problem.

Documentation is necessary to ensure that learning from the improvements is shared across the organization and with all workers involved. In order to have repeatability of changes by others, it is necessary to have documentation of the process through procedures or policies. The documentation becomes standard operation for all involved and ensures that improvements will be carried out in the same manner by all that operate the process.

Control charts are graphs used to analyze how a process changes over time. Data are plotted along a timeline. As depicted in Figure 2.14, a control chart always has a central line for the average, an upper line for the upper control limit, and a lower line for the lower control limit. These lines can be determined by desired outcomes or from historical data. By comparing current data to these lines, you can monitor to determine if the process variation is steady (in control) or is erratic (out of control).

Control charts are best suited for controlling continuous processes by identifying and correcting problems as they occur; when predicting the expected range of outcomes from a process; when determining whether a

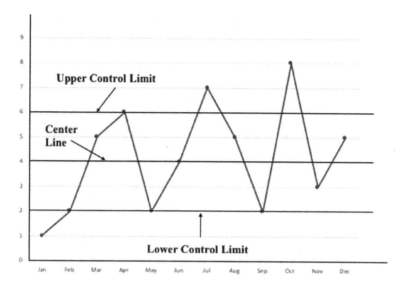

Figure 2.14 Example of a control chart. (Courtesy of PRESENTERMEDIA.)

process is stable or erratic; when analyzing patterns of process variation; or when determining whether your improvement actions should point to preventing specific problems or to making fundamental changes in the process.

The five-step action process for identifying a solution to eliminate waste has been explained along with multiple tools that support each of the five steps: identify, measure, analyze, improve, and sustain. Now that the tools have been shared along with examples of how to implement, it's time to use the skills learned. Start by conducting a walk-down in your facility/area, using your "Lean Lens" to identify a source of waste. Now apply the tools outlined in this chapter to improve the overall process. Remember to involve the workers most familiar with this process; they are the ones with the best ideas on how to improve and eliminate the waste.

2.5 *Assessing lean principles*

This section provides a framework to assess the implementation of Lean principles within any organization. It is important to understand how well your organization has implemented the five-step action process and learn what additional improvements can be made to achieve Peak Performance. Table 2.2 provides example questions that can be used when assessing programs and processes that have implemented or applied the five-step action process.

Table 2.2 Example Questions for Lean Assessment

Questions	Responses
1. Does the facility/project have a Strategic Plan for implementing a Lean Process?	
2. Do managers walk down their areas with a "Lean Lens" looking for potential areas of waste?	
3. Has training been provided to senior management on Lean principles?	
4. Once waste has been identified do managers develop a plan to identify, measure, analyze, improve, and sustain changes?	
5. Does the Management Team review progress toward actions to reduce the waste?	
6. Does the Management Team value the Lean process and encourage employees to get engaged?	
7. Are metrics/goals reviewed on a routine basis by the management team?	
8. Do Managers discuss opportunities to utilize Lean concepts?	
9. Are Managers managing for daily improvement, i.e., looking for incremental improvements? How does your team go about identifying these improvements?	
10. Are various tools utilized to reduce waste, i.e., fishbone diagrams, SIPOC, etc.?	
11. Is rework tracked and trended?	
12. Is standard work utilized? How do you verify standard work is used?	
13. Are the 5S principles utilized in the work area?	
14. What areas has Lean been implemented in (procedures, operations, work control)?	
15. Has a time-motion study been conducted to highlight areas where employees may have excessive waiting periods?	
16. Is work planned utilizing Lean concepts?	
17. Do your Managers encourage workers to make recommendations/suggestions?	
18. Are employees utilized as an important resource?	
19. Are the opportunities for Lean application identified, and communicated to the workforce?	

(Continued)

Table 2.2 (Continued) Example Questions for Lean Assessment

Questions	Responses
20. Are visual management tools utilized in the facility (production metrics, improvement actions, etc.)?	
21. Does the organization have a visible mission statement that employees know and identify with?	
22. Is the plant/facility meeting its goals? If not, why not?	
23. Do workers actively participate in the suggestions/idea program?	
24. Are metrics visible to the workforce?	
25. Are employees visibly engaged in reducing waste in the process?	
26. Are results from the Lean events published to the workforce?	
27. Do employees volunteer to participate in Lean activities to support elimination of waste?	
28. Do workers seek out the Management team to discuss opportunities to improve?	
29. Have the workers used Lean concepts and tools?	

References

1. Krafcik, John F. (1988). Triumph of the lean production system. *Sloan Management Review.* **30** (1): 41–52.
2. Daniel, T. Jones, and Roos, Daniel (1990). *The Machine That Changed the World.*
3. Simon, Kerri *SIPOC Diagram.* iSixSigma, Ridgefield, CT, Retrieved 2012-07-03.

chapter three

Principles of human performance and performance modes

3.1 Introduction

Let's begin by defining human performance. Human performance is simply behavior (B) plus results (R) equals performance (P) – B+R=P[1]. Behavior is what people do and say; it is observable actions. All behavior, whether bad or good, is reinforced by experiences and direct consequences. In a business environment, leaders are seeking to achieve Peak Performance, excellent production rates, zero work-related injuries, and outstanding customer satisfaction. Essentially, all industrial processes involve the operation of equipment, processes, or implementation of procedures by people, and we know that people make errors that can cause significant events.

Multiple studies indicate that the dominant cause of events is human error. The costs of events triggered by human error are undeniable and are frequently seen. If we could only prevent people from making errors, or at least anticipant when an error might occur, there would be an opportunity to correct and prevent the error, and thereby the resulting incident. An extensive European study of small business ventures concluded that a corporation that is focused on managing errors rather than preventing them is more likely to reduce its profitability[2]. Successfully managing human errors can yield the following benefits, among others:

- Fewer injuries and events
- Reduced event severity
- Reduced rework and downtime
- Greater efficiencies
- Reduced cost of poor-quality products
- A more motivated workforce

3.2 People are fallible

Studies have shown that managing human error throughout the life cycle of an operation will yield a positive return on investment. To effectively manage human error, and to achieve Peak Performance, we must encompass and understand several fundamental principles and beliefs of human performance[3,4]. Learning and integrating the principles documented in this chapter into leadership and worker practices will help build a foundation for understanding human behavior. Integrating these principles into day-to-day operations may require a change in standard operating procedures, not only for Managers and Supervisors but also for all employees.

Fallibility is an enduring characteristic of humans. Recognizing that people possess a wide range of capabilities, as well as shortfalls, is key to modifying behaviors and reducing errors. One of the most recognizable shortfalls of people is they will make mistakes; it's just human nature. No matter how smart an individual is, no matter how much training they receive, or how motivated an individual is, they can and will make errors. Even your *best* employees make mistakes, it's just human nature! An important question to ask yourself is: Do I want to rely solely on an individual to complete the most important steps in a process without some other control in place? Employees don't come to work seeking to fail. Error is not an intentional action; no one wants to purposely make errors.

Making an error on a particular task can be avoidable, notwithstanding the inevitability of human error.[5] Modifying work conditions to accommodate the existence of error traps when performing a task, or at least minimizing the chance for error, can prevent a significant event. Events caused by human error can signify a breakdown in the organization's process and procedures, not just an individual's actions. Opportunities for error in operations exist almost everywhere in a system or process. Safe operations are preserved by an organization's ability to recognize, adapt to, and counteract the disruptions and disturbances, including human error.

Human behavior observed within an organization is a direct reflection on that organizations' culture, leadership team, policies, practices, procedures, and processes. For example, if an employee follows a newly generated procedure, and as a result causes a fire because the procedure required adding an incorrect amount of a material, the incident should not be considered the employee's fault; instead the organization failed the employee.

All behavior, whether bad or good, is reinforced by experiences or direct consequences. Most workplace behavior is underpinned by the consequences the employee experiences when the behavior occurred. People tend to determine what is acceptable behavior and then model it. So, if the organization permits short cuts and trade-offs without any consequences, then employees will tend to view these behaviors as acceptable. These behaviors become ingrained in the way business is conducted and often lead to errors.

Depending on the reports and studies cited, error rates for humans can vary from 1 to 15 errors per hour. One study affirms that people, when performing any activity (except sleep) commit on average 3 to 4 errors per hour.[6,*] Typical human reliability can be on the order of 99.99% (1 error out of 1000 opportunities), depending on the task and the working environment[7]. The conclusion that one can make is that errors reduce the margin of safety and increase the likelihood of events. Not all errors have a bad outcome, some errors have led to great discoveries. For example, Dr. Spencer Silver, a Chemist at 3M, was trying to invent a superstrong adhesive, but unfortunately developed a low-tact one instead. Dr. Silver endorsed his new invention but no one could think of a good way to use it. Then a colleague, singing in his church choir, was having difficulty keeping his place in the hymnal. He remembered Dr. Silver's invention and tried some of the adhesive and it was the perfect solution. Dr. Silver's colleague recommended using the adhesive on small notes, and 3M launched the product "Press 'n' Peel." It was not wildly successful, therefore the product was tested again under a new name, "Post-It® Notes." Error is a behavior without consideration, it is not the result. Errors usually occur because of misinformation, inattention, or incorrect assumptions. Although the error that led to the discovery of Post-It Notes is a good story, often the results of errors are events that can set a company back many years.

3.2.1 Active and latent errors[10]

Errors generally fall into one of two categories – active and latent errors. An active error is an observable error – an action that changes the configuration of a piece of equipment, resulting in immediate adverse outcomes. Front-line workers are usually the ones that commit these types of errors because they are the ones that are hands-on with equipment operation and usage. An example of an active error would be manipulating an incorrect valve on a water system. An active error generates an instant undesirable consequence. These errors can go unnoticed at the time they occur and have no direct impact on personnel or equipment. Latent errors appear as weaknesses in processes and procedures, as inefficiencies, and unwelcome changes in values and practices. These mistakes are typically paper-based, such as incomplete procedures, policies, drawings, and design documents; latent errors occur as a result of workers unknowingly utilizing them. Whereas active error immediately appear, latent errors are not readily apparent and are typically made by administrative personnel and engineers, not front-line workers. Table 3.1[10] summarizes the attributes of each type of error. As you can see, latent errors are more difficult to identify and more challenging to manage than active errors.

* Also, Dr. Michael Frese of the University of Giessen and London Business School suggests that you make approximately three to four errors per hour on every task that you work on – the "Frese's Law of Error Frequency."

Table 3.1 Summary of Attributes for Active and Latent Errors

Questions	Active Error	Latent Error
Who makes these types of error?	Front-line workers	Administrative workers, procedure writers, Managers, support staff
Where does the error typically occur?	Equipment	Documents, values
When is it evident that an error occurred?	Immediately	Later, delayed
Is the error visible at the time it occurs?	Yes	No

3.3 Anatomy of an event[10]

Let's look at how events occur. Some refer to this as the anatomy of an event. As an example, Sally is a new operator at the Ruby Rabbit Company. She is tasked with flushing a tank with acid. This is a task that is only performed once or twice a year. Sally was requested to flush the tank with acid early in her shift. Her first task took a lot longer than expected, she became distracted, and forgot to flush the tank before leaving for lunch. After lunch she was assigned a different task, but before completing the task, she remembered that she had to flush the tank before the end of the shift, and it was getting late. Sally read the procedure and proceeded to flush the tank. Sally located the valve, checked the diagram in the procedure, and then opened the valve. But instead of the acid flowing into the tank as she thought it would, acid rapidly leaked onto the floor. An unexpected event has occurred. What happened, why did it happen, could it have been prevented? Figure 3.1 depicts the anatomy of an event.

3.3.1 Error precursors[10]

Let's break down the anatomy of an event into each type of potential failure mode. What are error precursors? An error precursor is an adverse prior circumstance at the job location that can increase the chance of making a mistake; it's a mistake waiting to happen. These circumstances typically occur when the requirements of a task exceed the ability of the employee or when the work settings exacerbate the boundaries of the employee's abilities to respond as expected[8]. These predictable mistakes or errors can be grouped into four categories: task requirements, work settings, employee competencies, and human characteristics. Table 3.2[10] provides some examples of each type of error precursor.

Chapter three: Principles of human performance   45

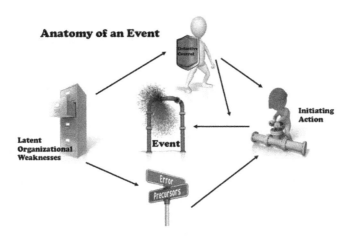

Figure 3.1 Anatomy of an event.

Table 3.2 Examples of Error Precursors

Task Requirements	Work Settings
Excessive workload	Distractions, disruptions
In a hurry (time pressure)	Unclear panels, or exhibits
Multi-tasking (a lot of tasks that need to be completed)	Equipment that is not functioning, finding an alternate way to complete the task
Reoccurring task (boring)	Concealed system responses
Irreversible actions	Variation from the standard, a change
Clarification of request	Unforeseen equipment status
Vague purposes, functions, accountabilities	No additional indication of status
Blurred criteria	Individual struggles
Employee Competencies	**Human Characteristics**
First time to complete activity, inexperienced	Anxiety/stress
Limited understanding of the task	Routine protocols
New method not utilized before	Theory, notions
Poor communications	Brashness, complacency
Deficient skills	Outlook, belief
Limited ability to resolve issues/problems	Imprecise view of risk
Haphazard attitude toward important actions	Preferences, conceptual short cuts
Low energy, tiredness, sickness	Reduced recall of information

3.3.1.1 Task requirement[10]

Task requirements are explicit mental and physical requirements needed to achieve a task that may exceed, or challenge, the ability of the employee. In addition, an "excessive workload" requires employees to sustain a high level of attentiveness. A perfect example is trying to complete a complex calculation while working in a noisy environment. "In a hurry (time pressure)" is the pressure or stress caused when trying to complete a task, and not having enough time to complete the task. "Multi-tasking" is performing several activities at the same time that require either mental or physical work, which may divide your attention. For example, writing an email while on the telephone. "Reoccurring tasks" are actions that are repeated multiple times during a shift or during a week, that are typically mundane or boring. Examples of reoccurring tasks include taking a reading on a piece of equipment, changing paper in a cash register, or putting fuel in a vehicle. "Irreversible actions" are actions completed where it is not evident as to how to reverse them or recover from them, without a significant time delay. Operating on the wrong part of a patient's body or tearing down the incorrect building are both examples of irreversible actions. "Clarification of request" is not really understanding the request or requirement, or having to interrupt what is being requested. An example to consider is repairing a motor, having the owner's manual, but not really being sure how the bolts are removed because the diagram is confusing, with "vague purposes, functions, and accountabilities." An example of vague purposes, functions, and accountabilities for the team includes working as part of a three-person crew to repair an assembly line. The crew consists of one electrician, one maintenance mechanic, and one operator. It is not clear who should perform which task to repair the line. A potential tool to use to provide clarity is swim-lanes (discussed in Chapter 2) as a way to clearly define who has what responsibility when multiple groups/disciplines are working together.

3.3.1.2 Employee competencies[10]

All employees are different, each has distinct skills and limitations, both mental and physical capacities that may or may not perfectly match a given task. How competent are employees the first time they are assigned to complete a task, or assigned an activity that has not been executed at the facility before? How familiar to the employee are expectations and performance criteria of the task? In these situations, how much knowledge or factual information about the task/activity does the employee possess? If the task or activity is new to the facility, will employees be familiar with the tools and/or methods needed to perform the job?

Communication skills are vital to ensuring tasks are completed as assigned and as outlined in procedures. Errors and unwanted events can occur when employees are unable to adequately communicate with their Supervisors and peers. Employees may exhibit a deficiency in skills or knowledge when a task is infrequently performed; they may have forgotten what the expected responses were or even how to operate the equipment.

In any facility, or as operational conditions change, something unexpected happens and employees must know how to respond. When employees are unfamiliar with the task or activity, employees are challenged in developing solutions and may be unable to manage the changing conditions. Another real challenge can occur when employees have a haphazard or complacent attitude toward important actions, such as not anticipating the hazards when completing a task. Often catastrophic events occur when employees are unaware of what is expected, conditions change, they have a haphazard attitude, are not able to resolve problems, or have low energy because of being sick or not feeling well.

3.3.1.3 Work settings[10]

General stimuli in the work setting, organizational structure, and cultural situations can affect individual behaviors. For example, distractions, unclear labeling, unforeseen equipment status, complex procedures, at-risk patterns and values, work-group mindsets toward hazards, work-control processes, and work-site environment such as lighting, noise, temperatures, etc., directly impact how employees work.

When an employee is working on an activity or task and is interrupted, which requires the activity to be paused or stopped, the employee can become unfocused and may not be as attentive as before to hazards in the workplace. If a worker's primary job is to monitor a control panel that consists of multiple viewing screens, a disruption can lead to confusion and may drive the worker to find alternate ways to complete a task due to frustration, resulting in incorrect actions.

Alternate ways to complete a task is when a worker did not complete an action as required because of an equipment failure, incorrect procedure, deficient material conditions, or maybe the employee was in a hurry and chose not to follow the defined work-practice procedure. Another example is when an action is taken by a worker and there is no response from the process or system; this is referred to as a concealed system response. When a concealed system response occurs, the worker is not aware that he/she is correctly operating the equipment. Utilizing alternative means to complete a task when there is no immediate feedback on what's occurring within a system can allow incorrectly completing a task to continue for a long time. However, at some point, the error may become evident as a result of a catastrophic event taking place.

3.3.1.4 Human characteristics[10]

Common traits, personality, and restraints may bias workers to err under adverse conditions. These traits include behaviors, short-term memory, stress, complacency, imprecise view of risk, outlook, and reduced recall of information. Stress can cause a worker to make a poor decision, lose attention and focus, and reduce their ability to recall key facts, which increase the potential to make an error. Human characteristics can lead to a worker ignoring changing conditions or process variances because of routine protocols, such as working on "auto-pilot." Additionally, workers can have a theory or notion without verifying facts, which may be based on perception of a recent experience and triggered by an inaccurate mental mode.

As workers become more experienced and continue to perform the same task over and over again, they can become complacent and overlook hazards, dangers, or assume the conditions will always be the same by not having adequate situational awareness. Preconceived beliefs can cause employees to become myopic and only see what the mind is wanting to see, missing information/data that was not expected or noticing one's own error. Evaluation of workplace hazards can also be dependent upon the perception of workers and their comfort in risk-taking, which can cause errors and limit the understanding of consequences. When employees encounter unfamiliar situations, some try to look for patterns and apply rules of thumb that can lead to conceptual short cuts to explain the differences. If employees are required to simultaneously deal with multiple data it may reduce the ability to recall necessary information, which can lead to errors.

3.3.2 Initiating event[10]

As we examine the scenario above with Sally as the new operator, the initiating event was having Sally open a valve to acid flush a tank. As we continue to explore Figure 3.1, what were some of the error precursors? Looking at task requirements, Sally may have felt pressured because of lack of time, and maybe even an excessive workload; she was definitely having to multi-task on her shift. In evaluating employee competencies, her Manager should have recognized she was a new operator, had never performed the task before, and her work skills were not mature because she was new to the position. What about the work setting? Did Sally receive any indication that the acid was *not* flowing into the tank as expected? Could Sally have been distracted because it was late in her shift? All of these factors could have played a role in the error that was made. Finally, Sally was probably feeling some stress to complete the task before the end of her shift, and she may have had an imprecise view of the risk involved in this task, such as acid flowing out of the tank instead of into the tank.

Chapter three: Principles of human performance   49

If you were the Manager of this group, knowing what error precursors are, you can address the precursors prior to starting the job through various means such as:

- Assigning a more experienced operator to work with Sally to perform the infrequent task
- Frequently observe task performance
- Speaking with employees about distractions and disruptions and the signs to be aware of that may indicate a lack of focus

As a Manager, you must know and understand the capability of each employee, facilitate a strong culture of reporting challenges, including personal ones, such as not feeling well or personal problems at home. Knowing and acting on these potential issues can make your operations much smoother and safer.

3.3.3   Latent organizational weakness[10]

As depicted in Figure 3.1, a latent organizational weakness is a concealed organizational weakness or flaw that lies inactive. These errors go unnoticed at the time they occur and have no immediate impact on personnel or equipment. Latent organizational errors appear as weaknesses in processes, procedures, as inefficiencies, and unwelcome changes in values and practices. These mistakes are generally rooted in paper-based actions, including procedures, policies, drawings, and design documents, and workers unknowingly utilize them. In our scenario with Sally, it is possible that the procedure she utilized to perform the task was incorrect.

3.3.4   Defective controls[10]

As depicted in Figure 3.1, defective controls are also known as shortcomings. Defective controls constrain the ability of protective measures to shield facility equipment, protect personnel against hazards, or stop active errors from occurring. Examples of defective controls include:

- Poor procedure steps
- Inoperable or mislabeled equipment
- Poor design or placement of equipment, including inaccessible areas

In the scenario above, it is possible the valves were incorrectly labeled, or the labels had faded, causing Sally to open the wrong valve in the system.

All of these failure modes can, and should, be predictable and manageable in any operation. The key to preventing errors and mistakes is to understand the underlying principles of how the error or mistake was made and then establishing defenses to prevent these mistakes. All human beings make mistakes, but we must expect, anticipate, and finally plan for these mistakes as we conduct our daily business.

3.4 Organizational culture

In any organization, we depend on employees to accomplish the task they are assigned to perform in support of making the business successful. We already know that humans make mistakes on average three to four times per hour, but as part of successful business operations we need to anticipate these mistakes, and understand how, and why, employees make these mistakes. Equally important is having an organizational culture that encourages reporting of errors or near misses; a mistake was made but there were no negative outcomes. Understanding where errors occur enables Management to review the process and accurately determine why so that future errors can be prevented. For example, Sally was a new operator and had an event that resulted in acid leaking from the operational process. What if Sally had quickly shut off the valve and then cleaned up the leak without ever telling anyone? What could be some of the potential impacts?

One impact may be that Sally accidentally touched the acid while cleaning up the leak and received burns to her hands. Another impact could be that Sally had cleaned up the leak with nearby rags and placed the saturated rags in a waste container, and another employee then emptied the trash and burned their hands. What if Sally never told anyone that the tank was not flushed with acid? Sometime later the tank is used, assumed to have been flushed, and another hazardous material is added to the tank that is not compatible with the acid, which caused an explosion that resulted in extensive damage to the facility and injured many workers. All of these examples are latent errors that could cause significant harm.

If we were to approach this event with a culture where people feel free to report errors, the outcome could be significantly different. For example, Sally, as a new operator, recognized something was wrong and immediately notified her Supervisor that acid had leaked onto the floor while filling the tank. The Supervisor responded to the event and began the conversation with Sally by thanking her for notifying him and asked if she needed medical attention. He then asked her to tell him what happened, what actions she took, where she was standing, and what response she expected. What were her next actions? The Supervisor may again ask Sally if she was hurt and needed medical attention. Wow, what a difference! The Supervisor now has the ability to have the event site properly barricaded, start an investigation, understands the configuration of the equipment (the tank has not been adequately flushed), and the leak can

Chapter three: Principles of human performance 51

be properly cleaned up. The real benefit is the learning as to what caused this event and how to prevent it from reoccurring. As Managers and Supervisors, we want and need employees to report *all* errors, even minor ones; that's how we learn and infuse the concept of continuous improvement in the minds and actions of workers, as well as continue to improve operations.

3.5 Human performance modes[10]

Human performance modes were developed to explain how employees conduct work according to the level of performance that appears to be sufficient to control the situation[8]. Information is processed by humans in one or more of three performance modes: skill-based (SB), rule-based (RB), and knowledge-based (KB). The performance mode is a function of the worker's awareness with the task, and the extent of attention required to complete the task. If a worker is very familiar with a task and routinely performs the task, for example, using a hammer, the worker is performing the task in a SB mode. If the work involves a task that is not familiar to the worker, uncertainty increases, and the worker will perform the task in the KB mode because they will need to focus their attention to spot critical information needed to perform the task. For example, what if an employee was asked to troubleshoot a compressor not properly running? The worker would be in the KB mode because the worker probably does not routinely conduct the task and is having to use diagnosis and problem-solving skills. When working in a KB mode, workers will look for ways to enhance their understanding of the situation in order to properly react; however, workers will default to the lowest level of mental effort they sense is needed to achieve the activity (avoiding any mental strain). As a result, workers may miss information needed for the situation, because of the perceived familiarity with the task. A worker's training, experience, and frequency of performing a task impacts the level of information processed. A worker may switch between levels of information processing while conducting a task, irrespective of the ease or the frequency of the task.

Let's look at a quick example to explain the three modes of performance. You decided to purchase a new rabbit from a breeder about 200 miles from your home. So, as you prepared for your trip to bring home your new pet, you loaded the car with a cage for the baby bunny, a water bottle, and a soft blanket. You jumped in your car to bring home your new pet, drove at the proper speed, and followed the map to the breeder's home; you were very familiar with both tasks, you were in SB mode. As you were driving to the location of the breeder, you encountered other drivers and situations on the road. As an experienced driver, you knew how to change lanes and what distance you needed to maintain from the other vehicles, based on rules you learned when you started driving. RB mode is described as: when situation (x) occurs, you either take or don't take some action (y). You were about halfway to the breeder's home and your electronic map application showed a delay of 10 minutes ahead of you that could easily delay your trip. Knowing you needed to arrive at the breeder's home on schedule, you pulled off the road and looked for an alternate route, hoping to arrive close to the time you told the breeder. You made a call to let the breeder know you may run late, and that you were taking an alternate route. You were required to use problem-solving skills to make a decision to find an alternate route and used your skills to accurately find a different travel route; this placed you in KB mode.

Using this example about traveling to pick up a new pet, consider that when you were driving to the breeder's home you were listening to music and were talking with a friend on your cell phone, what is the potential that you might have missed the information about a delay ahead on your map application? The likelihood of making an error is increased when trying to do more than one activity in one phase of information processing. This is why it is so vital to manage the environment that people work in to reduce the interruptions and distractions, and other activities that can adversely impact a worker's attention and abilities. Experienced and trained workers can knowingly focus on only two to three outlets of information (such as flow rate, motor speed, tank temperature) and still be effective[9].

3.5.1 Skill-based performance mode[10]

A better understanding of a worker's conscious and spontaneous behaviors as explained in the three performance modes – skill, rule, and knowledge – and recognizing the kinds of errors workers are inclined to make while performing the various modes, can be useful in driving for Peak Performance. Workers performing SB activities need sufficient tools to diminish slip-ups, and they need to be free of interruptions and distractions that erode focus, divide attention, and contribute to decline in memory that can cause an error.

The SB performance mode involves actions that have been performed multiple times, are predominately physical actions, and occur in a common setting. The actions are generally completed from memory without considerable conscious thought, as depicted in Figure 3.2[10]. When proficiencies are learned to the point of being routine, the burden on the working memory is typically reduced by 90%. A significant number of activities in a typical day are managed unconsciously by human nature, such as answering the phone, signing a check, or driving a car. Performing these tasks is an example of the SB mode.

Chapter three: Principles of human performance   53

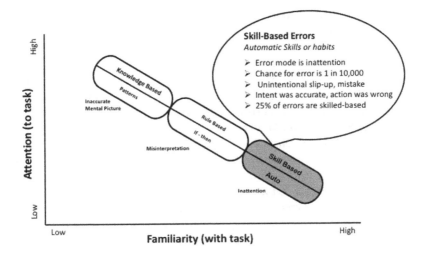

Figure 3.2 Skill-based mode.[10]

When a worker is in the skilled-base mode, they are able to function very efficiently by employing a pre-programmed string of behaviors that involve little conscious physical command. Table 3.3[10] lists examples of SB activities.

The primary error mode for SB performance is inattention; the worker knows what actions should be taken but makes an error because of a lapse in attention or focus. For example, John, who is a very experienced worker at the Ruby Rabbit Company routinely flushes the lines for both pellet machines, System A and System B.

John was asked to turn off the water to System A, but inadvertently turned off water to System B. John knows the systems well and routinely performs the task, so why did he make a mistake? John could have been distracted because of loud noises, others talking, or maybe was daydreaming about a party that evening. In any case, John lost focus for some reason, allowing the error to occur. John had an accurate understanding of the task and had the correct intention, but just made a mistake.

Table 3.3 Examples of Skill-Based Activities

Using a screwdriver or other hand-held tool	Performance repetitive calculations
Washing the car	Fueling a vehicle
Opening a valve	Replacing the filter on a refrigerator
Taking readings on equipment	Calibrating a piece of equipment
Controlling flowrate	Analyzing route samples
Reading a book	Shoveling snow

Another challenge with SB activities is that workers are accustomed to conducting the task, causing the perceived risk to be less than what the actual risk is, which can lead to overconfidence, and potentially becoming insensitive to the actual hazards. Approximately 90% of a worker's tasks are performed in the SB mode, with an error rate of 1 in 10,000; for every 10,000 times a worker performs a task, they will make an error. The best way to prevent an error is to recognize that workers performing tasks in the SB mode make errors because of inattention. Manager and Supervisors should look at the work environment and identify any potential distractions. When looking at errors that occur in the workplace, only 25% of errors occur in the human performance SB mode.

An error has occurred and the employee was in the SB mode; what are some potential corrective actions? Since the worker knew what was expected, retraining is not an appropriate action, but maybe use of a checklist to limit the number of steps performed based on memory is an appropriate corrective action. Another option is to have a peer check the steps to ensure they are correctly performed, or only having two tasks going on at one time. Table 3.4[10] lists examples of corrective actions for preventing errors performed in the skill-base mode.

3.5.2 Rule-based performance mode[10]

Workers shift to the RB performance mode when they observe a need to revise their pre-programmed behaviors because they have to take into account changes or unexpected conditions. The work condition has altered such that the proceeding activity (skill) no longer applies. This activity is typically one that workers have confronted before, or have been trained to handle, or have a procedure that addresses the unexpected condition.

Table 3.4 Examples of Corrective Actions for Errors Made in Skill-Based Mode

Review previous steps in a procedure if interrupted or distracted before continuing	Use caution statements in procedures
Utilize self-check before continuing to ensure proper actions are taken	Conduct pre-job briefing to ensure all workers understand the expectations and ensure everyone's head is in the game
Use peer-check to ensure steps are completed as required	Reduce distractions in the workplace, for example cell phones, music, etc.
Practice or use mock-ups	Limit task conducted at the same time to only two
Use awareness tools or reminders such as signs	Add distinctive labels for similar or look-alike equipment

Chapter three: Principles of human performance 55

It is called the RB mode because workers utilize memorization or written rules. Table 3.5[10] lists examples of RB activities. These rules may have been acquired as a result of previous interactions with equipment or the facility, through formal training, or by working with experienced workers.

The level of conscious command is somewhere between that of the knowledge- and SB modes. The RB level follows an if (symptom X), then (situation Y) logic. The objective in RB performance is to expand one's interpretation of the work condition so that the correct response is selected and used. Documented procedures, which include anticipated conditions or potential event scenarios, drive employees to perform work in the RB mode. Procedures include predetermined solutions to possible work conditions that require specific responses. Not all activities directed by a procedure are automatically RB performed. In normal work conditions, such tasks are commonly SB for the experienced user.

Errors that occur when employees are working in RB performance may be corrected through retraining. Generally, the worker has misinterpreted a requirement or a "rule." The worker has applied an incorrect rule to a given function; or, conversely, has used a correct rule in a wrong function. Table 3.6[10] identifies actions that may be applied to correct or prevent errors made in the RB mode.

Table 3.5 Examples of Rule-Based Activities

Deciding whether to replace a water valve during a preventive maintenance inspection	Performing checks for quality control
Responding to a low-pressure alarm	Using emergency response checklist
Estimating the change in tank level based on pressure change	Developing work packages and procedure
Listening to equipment for excessive noise (vibration) during operator surveillance in operational areas	Review of financial balance sheets

Table 3.6 Examples of Corrective Actions for Errors Made in Rule-Based Mode

Organizing work specialty units (e.g., technical specialist, component engineering, work planners, etc.)	Run through transition between procedures before performing the task
Having Supervisors observe workers using checklist and check off steps, and repeating back important steps	Clarification of unclear requirements, such as explaining "how to" do a mandatory action or explaining the expected outcome to determine whether success was attained
Provide training on systems or important steps in a task	Practice verbalizing the action intended to be taken.

RB errors can be recognized or mitigated by individuals demonstrating a questioning attitude, by calling a time out, or by stopping work when they are unsure. Peer checks can also be used to stop someone from committing a substantial error.

Since RB activities require interpretation using an if–then logic, the predominant error is misinterpretation. Workers may not fully comprehend or notice the equipment or facility conditions required for a particular response. Human error involves deviating from an approved procedure, applying the wrong response to a work condition, or applying the correct procedure to the wrong condition. Errors are also sometimes due to inadequate recollection of procedures. For example, an operator not recalling the correct sequence when performing a procedure.

RB and KB performance modes encompass workers making choices. As depicted in Figure 3.3[10], when workers are less familiar with an activity, the chance for error increases to roughly 1 and 1,000, and roughly 60% of all errors are RB. Example actions to reduce errors while working in the RB mode include procedure adherence, pre-job briefing, questioning attitude, peer-checking, and concurrent verification, to name a few.

Let's look at an example of working in a RB performance mode. Joe is an operator at the Ruby Rabbit Company. Joe has 30 years of experience making bunny pellets at the Ruby Rabbit Company. He had a two-week cruise planned over the fourth of July holiday. Joe felt good about taking off over the fourth because the plant shuts down for routine maintenance and typically only performs minor upgrades. Joe began his vacation as planned and really enjoyed the time with his family, saw some beautiful

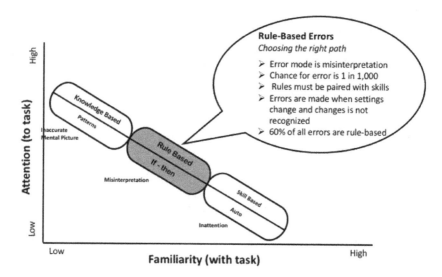

Figure 3.3 Rule-based mode.[10]

Chapter three: Principles of human performance 57

sights, and got some much-needed relaxation. When he returned to work on Monday, his first job was to restart the pellet food line as he normally did after each plant shut-down. Joe and his team had their typical morning briefing about the changes made to the plant during the shut-down; however, Joe's mind was still on his vacation. Joe started the pump to begin adding water to the tank to start cooking oats for the pellets; the pump light came on as normal, a good sign everything was working fine. After adding water to the tank, Joe hit the stop button on the pump and the indicator light went out. Joe went on to complete the other steps in the procedure. About 20 minutes later, the high-level alarm on the tank sounded. Joe acknowledged the alarm, assuming it was the typical malfunction of the cooking system. The high-level alarms had gone off in the past, but it was because some of the oats had boiled higher than normal, but nothing to worry about. Joe went back to completing the rest of his procedure. About 15 minutes later, Joe heard a loud sound and people screaming. Joe stopped the procedure he was working on and ran to find his best friend covered in oats and water. He asked his friend if he was hurt and what had happened. The oat tank had overfilled causing the piping to disconnect from the tank and spill the contents on the floor. When Joe's coworker walked by the tank he slipped on the water and oats and fell and broke his leg.

What performance mode(s) was Joe working in? To begin with, Joe was working in the skilled-based performance mode; he had routinely performed the cooking procedure many times. However, his performance shifted into a RB performance mode after he thought he shut off the water to the oat tank. He just assumed that the system worked properly. Even when he received a high-level alarm on the tank, he ignored the indicator, assuming the system was properly operating.

The Ruby Rabbit Company conducted a comprehensive root-cause analysis of the event to determine what happened, and how to prevent the same event from reoccurring. The root cause of the event was that the pump did not shut off when the button was pushed. During the two-week shut-down, the pump was replaced and incorrectly wired. If Joe had not been distracted when the morning briefing was conducted, he would have heard that the pump was replaced and the high-level alarm was repaired during the two-week shutdown. Joe assumed that because he pushed the stop button on the pump, the pump stopping pumping water into the oat tank. Joe also ignored the high-level alarm, again just assuming it was the typical malfunction of the high-level alarm. Both of these errors occurred during the RB performance mode: if X, then Y occurs. If Joe pushed the stop button on the pump (X), then the pump would shut off (Y). Joe also ignored other indicators, for example the high-level alarm on the oat tank, and it is also possible that Joe could have heard the pump running after he thought he had shut the pump off.

Knowing that Joe was in RB performance mode, what type of corrective actions might be appropriate to prevent this event from happening again? Joe could be retrained to ensure he understands other indicators on the system, which would help him recognize environmental and workplace changes. Joe could also have a coworker peer check that the pump is shut off after pushing the button or have his Supervisor discuss the other indicators on the system that can be utilized as verification that the pump is working as designed.

Understanding requirements, and knowing where and under what instances those requirements apply, is cognitive in nature and must be learned or gained in some way. Many work-related events have occurred because employees did not recognize that the original task had changed because of failure in human-performance modes, such as the transition from routine maintenance to troubleshooting the system.

3.5.3 Knowledge-based performance mode[10]

Workers shift to a KB performance mode when the cause of the issue cannot be ascertained by applying any accessible rule; the worker may be forced into KB performance level. Table 3.7[10] lists several examples of KB performance mode activities. The first approach in responding to a task is to find a similarity between the unfamiliar conditions and some pattern of events for which the worker has rules (RB performance mode). If a rule can be identified that applies, the worker will lapse back to RB performance and utilize the suitable action rule. However, if an appropriate related rule cannot be identified, it may require the use of other knowledge (technical, engineering, etc.) to manage the incident.

The challenge with KB errors is that other factors are important when developing corrective actions, such as the demand on the information-processing skills of the worker when a condition has to be evaluated. It is not unexpected that workers do not perform well under high stress, or when encountering unfamiliar conditions where they are obliged to "think on their feet," in particular when there is a lack of rules, routines, and procedures to manage the condition.

Table 3.7 Examples of Knowledge-Based Activities

Troubleshooting a heating system	Conducting meetings to tackle problems
Performing a design evaluation	Conducting root-cause analysis of incidents
Reviewing a procedure for "intent of alternation"	Performing trend analysis
Sorting out conflicting indicator boards	Allocating resources

Chapter three: Principles of human performance

As depicted in Figure 3.4, errors generally result from an inaccurate mental picture when using the KB performance mode. The challenge to the worker is in reacting to a different condition than what was expected, and the level of uncertainty may be high. When operating in a KB performance mode the error rate is 15%, and the chance for error is 1 to 2 in 10.

Corrective actions to reduce KB performance errors are more complex. An analysis of what went wrong is necessary to ensure that the corrective action(s) really address the cause of the event. Table 3.8[10] presents several examples of corrective actions associated with errors that occur when operating in the KB performance mode.

There are several scenarios that could have led to an event. For example, the worker's understanding and knowledge of the system, principles, or theory was insufficient to prevent the event from occurring. It could be that the worker's knowledge was acceptable, but other workers were

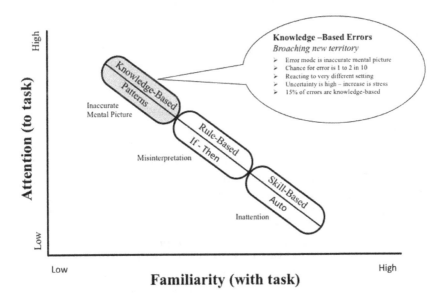

Figure 3.4 Knowledge-based mode.[10]

Table 3.8 Examples of Corrective Actions for Errors Made in Knowledge-Based Mode

Enhancing problem solving proficiencies/root-cause analysis	Appoint a "Devil's Advocate" during crucial decision-making meetings
Improve awareness of workflow processes	Avoid complacency
Enhance communications	Discussion and exchanging of ideas

deficient in problem-solving skills, or the group was in "passenger mode." When a work group performs in "passenger mode," everyone is not independently thinking, or the group has failed to properly communicate the resolution of the issue. Additionally, the work group may not be capable of making decisions in an emergency situation. Another type of behavior that could impact performance is called "vagabonding," where an overburdened worker devotes their attention to one cursory problem after another, without resolving any of them.

So, what are some corrective actions for KB errors? Training or retraining may help, and proactive coaching is always a good solution to helping workers eliminate errors in any performance mode. Peer evaluations have been found to be particularly useful for addressing errors that occur because of the KB performance mode.

Let's look at an example of a KB performance mode error. Bob and two other workers at the Ruby Rabbit Company were working the night shift to make rabbit treats. Bob was the senior operator on the crew, Bill had been with the company for about 5 years, and Robin had been with the company for about 18 months. All three workers had been trained on making rabbit treats and had been working together for about a year on the shift. It was Thursday night, the last night of the week, and the crew had produced a lot of rabbit treats. The Supervisor held the nightly brief as usual, but he stressed that they need to produce about 1,000 more treats on their shift because the day shift was behind on production. The day shift did make a batch of rabbit treats but did not have the time to press and bake them. Bob was complaining that the day shift was always behind on production numbers and resented the fact that the night shift *always* had to take up the slack, and his team-mates agreed. Bob read his procedure to start up the press for the making the rabbit treats while the other two read the procedure for starting the ovens. Bob completed his task and went back to the control room to wait for the pressure to increase on the press machines so he could start making the treats. Bill and Robin were engaged in starting up the steam plant to begin heating the ovens. Once the stream plant was online, Robin turned the valve to initiate steam flowing into the ovens. Bob was in the control room watching the temperature gauges on the control panel, waiting for the ovens to heat up enough to start baking the treats. Bill and Robin reported back to the control room after starting up the steam plant and initiating steam to heat up the ovens. About 30 minutes elapsed and Robin commented that it was taking a long time for the ovens to heat up, wondering if something was wrong. Bob commented that is was cool that night and sometimes it takes a little longer. Another 15 minutes passed, and Bill stated that he thought they should go and investigate why the oven was taking so long to heat up. Again Bob told him to just give it a few more minutes. About 15 minutes later, the high-temperature alarm on the ovens sounded; all three of the workers looked at each other in dismay, why was that alarm sounding? The temperature on the ovens did not reflect the required temperature needed to bake the treats. Both Bill and Robin looked to Bob for guidance, since he was the senior operator on shift; Bob said it was probably a false alarm on the ovens. Bob pointed to the temperature gauge on the control panel, and it still did not show the ovens were at the correct temperature. Robin asked if it was possible that the gauge was faulty. Bob responded no, of course not, it had never failed in all the years he had worked at the Ruby Rabbit Company. Not too long after that, an explosion was heard in the plant and the three workers went to see what had happened. The ovens were on fire, but more importantly, the workers were confused as to what to do next because the fire had spread to other pieces of equipment. Bill and Robin were waiting on Bob to provide directions, not thinking about what they could do to help, just looking to Bob for direction. Bob was confused and starting to panic because he had never been involved with such a significant accident.

What errors were made in this example? Both Bill and Robin were relying on Bob. Bob was confident in his understanding of the ovens heating up, not really questioning the fact that the temperature gauge could have failed. Both Bill and Robin were in "passenger mode," not independently thinking or investigating why the ovens were not reaching the appropriate temperature in a timely manner, and, when the explosion occurred, again relying on Bob for guidance. Bob was frantic and not resolving any of the issues. The three workers did not try to solve the problem as a team, exchange ideas, or really even communicate with each other. After the incident was over, the Ruby Rabbit Company completed a comprehensive root-cause analysis to enable them to develop the proper corrective actions to prevent reoccurrence of the event. The situation provided was an example of a KB performance error mode.

In this chapter, you were introduced to the principle of human performance. All humans make mistakes, it's just human nature. We typically make three to four errors every hour. Even the best performers make mistakes because people are fallible. It is necessary to have a clear understanding of why it is important to have a strong organizational culture, one that builds on trust, engaged employees, and the willingness to learn and continue to improve. Errors can either be active or latent. Error-likely conditions can be predicted and, therefore, strategies can be introduced to eliminate the errors or reduce the impacts. Individuals work in three performance modes: skill-based, rule-based, and knowledge-based. Understanding how workers perform, and associated performance modes, can lead to the development of corrective actions to prevent reoccurrence.

3.6 Assessing Human Performance Principles

This section provides a framework to assess the implementation of Human Performance principles within any organization. It is important to understand how well your organization has implemented human performance tools and learned what additional improvements can be made to achieve Peak Performance. Table 3.9 provides example questions that can be used when assessing programs and processes that have implemented.

Table 3.9 Assessment for Human Performance

Question
1. Any visible management support for human performance process?
2. Does the plant/facility have a Strategic Plan to reinforce the implementation of human performance principles?
3. Was human performance training provided to the workforce both initial and refresher training?
4. Are human performance metrics maintained and trends to make mid-course corrections?
5. Any evidence that human performance is included as part of work planning?
6. Any evidence that human performance is included as part of post job reviews?
7. Is feedback solicited and acted upon?
8. Are metrics and trends discussed with the workers to support improvement?
9. Is the management team visible in the plant/facility coaching the workers to reinforce expectations?
10. Did the management team discuss error-likely situations with the workers?
11. Are lessons learned discussed with the workers?
12. Are critical steps reviewed with the shift workers prior to starting a job or a specific task?
13. Do workers discuss critical steps among themselves looking for lessons learned from co-workers?
14. Are self-assessments conducted to review the implementation of human performance tools?
15. Did the organization display a questioning attitude, or interest in learning and getting better, improving efficiency, and overall production?
16. Does the organization conduct benchmarking activities with other companies to identify best practices?
17. Does the Management Team observe work activities and behaviors in the facilities to gain a better understanding of how work is actually accomplished?

(*Continued*)

Table 3.9 (Continued) Assessment for Human Performance

Question
18. Does the organization utilize independent oversight as a way to uncover potential blind spots?
19. Does the organization utilize a change management process before implementing significant changes (such as large organizational changes, major changes in operations, etc) to limit the overall impact to the workers?
20. Does the organization report errors and near misses to improve performance?
21. Do employees feel free to bring up issues to the management team?
22. Do employees believe that if they bring up an issue that the management team would act to resolve it?
23. Are a task previews conducted?
24. Are a pre-job briefing conducted?
25. Are critical steps discussed during the pre-job brief?
26. Are error-likely situations identified/discussed?
27. Are workers following written procedures?
28. Are workers utilizing good communication techniques during the performance of tasks?
29. Are issues openly discussed and followed up on?
30. Are post job reviews held after task are completed to identify issues and are lessons learned added to the work package/checklist before the task is performed again?
31. When workers are required to work on similar or look alike components does the team use flagging techniques to ensure the proper piece of equipment is worked on?

References

1. Van Dyck, C., Frese, M., and Sonnentag, S. (2005). Organizational error management culture and its impact on performance: A two-study replication. *Journal of Applied Psychology*.
2. Institute of Nuclear Power Operations (1991). An Analysis of 1990 Significant Events (INPO 91–018). And, a follow-up report from the Human Performance Initiative Special Review Committee, *Recommendations for Human Performance Improvements in the U.S. Nuclear Utility Industry* (November 1991). INPO, Atlanta, GA.
3. Adapted from the Institute of Nuclear Power Operations (September 1997). *Excellence in Human Performance: Building on the Principles for Enhancing Professionalism (handbook)*. INPO, Atlanta, GA, p. 3.

4. Center for Chemical Process Safety (1994). *Guidelines for Preventing Human Error in Process Safety*. American Institute of Chemical Engineers, New York, NY, pp. 12–17, and 103–107.
5. Spear, S. (2009). *The High Velocity Edge*. McGraw-Hill, New York, p. 46.
6. Civil Aviation Safety Authority of Australia (2013). *Safety Behaviors Human Factor: Resource Guide for Engineers*. Civil Aviation Safety Authority, Canberra, Australia.
7. U. S. Nuclear Regulatory Commission (1983). *Handbook of Human Reliability Analysis with Emphasis on Nuclear Power Plant Applications, Final Report* (NUREG/CR-1278). Prepared by Drs. A. Swain and H. Guttman Sandia National Laboratory. pp. 20–13.
8. Reason, J. (1998). *Managing the Risk of Organizational Accidents*, Ashgate Publisher, Farnham, UK. pp. 68–70, and p. 142.
9. Reason, J. (1990). *Human Error*, Cambridge University Press, Cambridge, UK, pp. 53–55.
10. U.S. Department of Energy, (2009) *Human Performance Improvement Hand Book*. Published by the United States Department of Energy, Washington, DC.

chapter four

Human error defenses

4.1 Introduction

This chapter describes tools that can be utilized by both Managers and employees to reduce the chance of error and/or reduce the impact of an error. The tools and techniques in this chapter will help anticipate, prevent, and capture active errors before they can cause damage, and assist in eliminating latent errors. The tools are intended to be utilized within an organization with a strong culture of reporting near misses, learning from mistakes, and a culture that instills trust among employees.

4.2 Human error prevention[2]

Human performance tools are routinely used when performing work, regardless of the task's risk or complexity. Each tool is described using the following criteria:

- Background: practical information about the tool's purpose and prospective boundaries
- When to utilize the tool: indicators for when the tool could be used
- At-risk practices to prevent: a set of behaviors, attitudes, assumptions, or circumstances that are inclined to reduce the effectiveness of the tool

The purpose of the human performance tools described in this chapter is to help Management and the worker gain and sustain definite control of a work condition – that is, what is envisioned to happen actually happens. Another way of saying this is: "Do the job right the first time." Before completing an action, a worker comprehends the significance of the action and its intended result. Every human performance tool will initially slow you down before ultimately speeding a worker up by preventing delays caused by active errors. When used carefully, human performance tools give workers more time to think about the action that they are about to complete, what is about to happen, and what they can do if things do not go as predicted. Excellent workers know intuitively what it takes to be successful; workers know they must spend time to think before taking any action, in order to get it right the first time. The goal of workers should be to ensure error-free performance, especially at critical steps in the process. Learning and understanding human performance tools will support a reduction in errors.

These tools apply to all forms of human performance irrespective of the type of work performed. Workers can, and will, make mistakes regardless of how prudently they use human performance tools. It is part of being human. Workers typically sense considerable pressure to perform their work flawlessly, either because of the pressures of production, such as making set goals and/or limited resources. The use of human performance tools do not ensure perfect performance, but workers can significantly reduce their chances of errors by using the tools carefully and rigorously. Because workers will still make mistakes, Supervisors can improve the chances of operating event-free by utilizing a variety of overlying defenses, barriers, and controls, while monitoring their effectiveness. Together, workers and Supervisors can minimize the probability of an event by reducing the occurrences of active errors and effectively managing defenses.

Fundamental human performance tools should be routinely used, whenever needed, without being prompted by a Supervisor. If carefully used, they serve as a basis for everyday successful performance. If workers apply these tools during work activities, it will help them to maintain positive control, regardless of their view of risk.

A worker's mindset toward the task sets the platform for success. Situational awareness tools assist the worker in having an accurate picture of equipment and workplace conditions, including any hazards, and the possible consequences of a mistake. These tools help workers ensure critical steps are performed accurately. A critical step is one that is not easily reversed and typically has significant consequences if incorrectly performed; it is also called the "point of no return." Self-checking helps a worker remove any doubt as to what action is required to be performed, when, and on what item. Whenever changing a component's setting or recording data, the worker self-checks the step. When a procedure or other guidance document is used, expectations exist for following the procedure as written. If verbally communicating information important for safety or production, workers should use effective communication techniques to verify understanding before acting. Together, these essential tools help the worker reduce human error, when used correctly[2].

4.2.1 Task preview[2]

A task preview helps the worker to understand the steps of an activity to enable error-free performance – again, doing the job right the first time. Before participating in a pre-job discussion or starting work, workers review procedures and other related documents to acquaint themselves with the breadth of work, sequence of actions, and critical steps. Workers

Chapter four: Human error defenses

may want to review the work site or talk with other workers who might have previously performed the job. The task preview also helps the workers contemplate how their actions could affect safety and production, and motivates conversations with other workers about human performance. It provides an organized, risk-based appraisal of the work activities from a human performance perspective and improves the worker's situational awareness while at the job site.

It is always recommended to conduct a task preview. During the task preview the worker identifies the critical steps, likely errors for each one, and potential consequences. It is also useful to consider the "worst thing that could happen." Once the errors and consequences are identified, applicable human performance tools can be implemented. Workers should understand which human performance tools to use and why. During the task preview, it is good practice to review feedback from others who have previously conducted the same task or something very similar, for lessons learned. While considering previous experience, workers should think about how they can prevent similar errors.

4.2.1.1 When to utilize the tool[2]
- Before starting a new task
- Before attending any pre-job discussion
- Just before performing a critical step
- After extended pause in an activity (i.e., started the job several days ago, and now going to complete the task)

4.2.1.2 At-risk practices to consider preventing[2]
- Not providing workers time to review procedures/work documents
- Workers not being prepared for a task
- Conducting a generic discussion on human performance tools versus being specific about which tools are most useful
- Neglecting to discuss specific human performance tools to use for each critical step
- Concerns of the workers not openly discussed
- Not reviewing lessons learned from previous tasks

4.2.2 Job-site review[2]

The goal of a job-site review is to improve the situational awareness of a worker when they first arrive at the work site. This tool has also been referred to as "Take Two for Safety," or in other words, stop, look, and walk down the job site where you plan to perform the work, identify and assess any existing hazards, and then ask yourself if you will be creating any hazards as a result of completing the task. A well-performed job-site review may take longer than two minutes, but workers should take as much time as needed to develop an accurate understanding of critical factors, plant/equipment condition, the work environment, hazards, and even potential impact to other workers. Utilizing this tool will help workers become familiar with the work area, enhance their questioning attitude, and improve their situational awareness techniques.

4.2.2.1 When to utilize the tool[2]
- Upon reaching the actual job site
- Prior to making contact with important equipment from a risk perspective
- During a review of a work package
- When a potential safety hazard exists
- After extended breaks or pauses, such as lunch

4.2.2.2 At-risk practices to consider preventing[2]
- Rushing, not taking the time necessary to look around the entire job site
- Thinking that "routine" or a "simple" task means limited risk or no risk
- Believing nothing can go wrong or nothing bad can happen
- Not talking about potential hazards or defenses with coworkers
- Not expressing "gut feelings," "pit in your stomach," or uneasiness you are feeling

4.2.3 Questioning attitude[2]

Attitude can be defined as a state of mind or a feeling toward a subject or object of interest. A questioning attitude promotes good situational awareness and boosts thinking about safety before actions are taken. Being aware of the work environment helps a worker maintain a precise understanding of work conditions at a given time, preventing blind spots, such as overlooking the obvious. This tool can help workers notice immediate hazards, warning signs, and issues with the work environment or the work planned. A questioning attitude can cause the worker to stop and resolve hazards, warnings, or issues before starting the task. When faced with uncertainty, workers must search for the facts, not assumptions, to divulge the truth about the situation, which will eliminate the doubt.

Complacency and lack of information weakens awareness. A questioning attitude encourages an inclination toward facts, over assumptions and opinion. Workers tend to assume things are right and the task will always go as planned. Workers are unwilling to fear the worst, in general, and a strong questioning attitude can overwhelm the desire to justify away the "gut feelings," and encourage a belief that something may not be right. When workers ask questions such as "What if …," or "Why?" it

Chapter four: Human error defenses 69

supports improving identification of incorrect assumptions and possible mistakes. Facts are an essential attribute that can be demonstrated and are observable behaviors and/or evidence. Without adequate facts, the worker should stop the task to address the erratic work condition that could lead to either a serious error or a significant incident.

A good pre-job discussion promotes a questioning attitude. A pre-job discussion can inform workers of what should be expected and what should not be expected. An informed worker understands the potential hazards, critical steps, and error-likely situations as well as the potential outcomes before starting the work activity[2].

4.2.3.1 When to utilize the tool[2]

The following warning signals can provide a chance to resolve doubt before continuing with a task:

- During self-checking, making sure you are ready to conduct the task
- Before conducting an important step or part of an activity
- When making a determination about an important task
- When having doubt, confusion, or reservation about an activity
- When having a "gut feeling" that something might not be right
- When faced with unexpected changes in conditions
- When inconsistencies occur between procedures, work package, and/or actual work-site conditions
- Following the discovery of unanticipated results
- Learning about missing information or resources
- After hearing those dangerous words from another workers: "I assume," "maybe," "should be," "not sure," "might be," "we've always done it that way," "probably," and/or "I think"

4.2.3.2 At-risk practices to consider preventing[2]

- Not stopping occasionally (taking a timeout) to verify the workers understanding of the work conditions
- Continuing with a task when questions arise
- Expecting nothing can go wrong
- Trusting the task is "routine" or "simple" means limited risk or no risk
- Making assumptions
- Attempting to make the reality of the work conditions match the expectations (mental model) rather than seeing what is actually there
- Believing the first thing that comes to mind, accepting the initial impressions, or taking assessments to be accurate
- Disregarding understated changes or slight discrepancies
- Letting emotions guide decisions rather than objective evidence
- Believing objective evidence without questioning its legitimacy

4.2.4 Pause when unsure or uncertain[2]

When you are in an unfamiliar area, things don't feel right, you feel challenged, uncertain, or muddled. Thinking back to the previous chapter, this is when you are in a knowledge-based performance mode, where the error rate is high (1 in 2 to 1 in 10). The best action to take when you are unsure or uncertain is to pause and ask others for support. Pausing when unsure of what action to take reminds workers to obtain more accurate information about the work condition before proceeding with the activity. It requires a short break from working on the task to enable workers, their Supervisor, or other knowledgeable individuals to converse and determine the appropriate actions to take before resuming the task. If a question arises and it is not clear how to resolve it, stop and ask for help! Everything is not always as it seems.

4.2.4.1 When to utilize the tool[2]
- When you have a feeling of uneasiness, doubt, confusion, or persistent questions
- When discovering situations that are not consistent with the procedure or other work documents
- When work is outside of the scope of the plan or procedure
- When unanticipated results or unfamiliar conditions are met
- When the expected action does not happen
- When lacking knowledge of a task to be completed
- If a peer states doubt or concern

4.2.4.2 At-risk practices to consider preventing[2]
- Disregarding conflicting evidence or points of view
- Overlooking the concerns of less-experienced workers
- Refusing to seek support from more-experienced/-knowledgeable workers
- Not seeking support for fear of shame
- Feeling insufficient if you seek support
- Assuming the activity is a "routine" or "simple" task
- Assuming nothing can go wrong
- Believing "skill of the craft" is adequate to deal with the condition
- Not having a well-defined terminate condition

4.2.5 Self-checking[2]

Self-checking helps the worker focus their attention on the appropriate process or piece of equipment. This check encourages the worker to think about the actions needed to complete the task, comprehend the expected outcomes (i.e., what should the worker expect to happen after they open

Chapter four: Human error defenses 71

the valve), and then verify the results of what was expected after the actions were taken. This tool, when carefully and thoughtfully used, can improve attention and thinking before a physical action is executed. An example of how this tool can be used is presented below.

John who works at the Ruby Rabbit Company was asked to fill a tank with water. John thinks about which valve to open to fill the tank, he locates the physical valve, he also looks at the drain valve on the tank to ensure it is closed. He knows what the expected outcome is when he opens the valve – the tank level will rise, and he can verify this result by using the sight-glass on the tank. Then John thinks about whether the proposed action is the right action for the task he was asked to complete. If John is not sure about the actions he should take, he should resolve any questions before proceeding. After John opens the valve and starts filling the tank with water, he pauses to review the results of his actions to decide if the right results were achieved. This tool works well if the worker has a good technical understanding of the system and knows the appropriate action to take.

Self-checking is predominantly effective for skill-based, repetitive tasks, which workers usually accomplish without a lot of conscious thought. When changing a component's status, (i.e., opening or closing a valve), attention must be focused on when the risk is the greatest. Rigor and care are essential when utilizing the self-checking tool. This tool can also help prevent errors when recording or entering data and performing calculations[2].

4.2.5.1 When to utilize the tool[2]
- When operating or changing plant equipment or controls
- When entering data into a computer or logging it on a form
- When doing a calculation
- When amending procedures or drawings using cut-and-paste on a computer or by making handwritten comments
- When collecting components that have similar parts that could possibly be switched

4.2.5.2 At-risk practices to consider preventing[2]
- Self-checking without referring to the guiding procedure or document
- Executing several manual steps in quick succession
- Executing more than one step at a time
- Executing a step when doubts or inconsistencies exist
- Not understanding the intent of the procedure step before executing it; utilizing a procedure without comprehending it
- Talking with another worker while executing a task
- Focusing attention on something other than the equipment being operated

- Losing visual or physical contact and not self-checking before continuing
- Not understanding if the step is a critical step
- Being sleepy or fatigued while executing a critical step
- Choosing to not take the time necessary to verify the results are correct

4.2.6   Procedure use and compliance[2]

Procedure compliance is grasping the procedure's intent and purpose and following it. The worker executes all steps as written in the sequence stated by the procedure or document. If the procedure or document cannot be completed as written, then the task is paused, and the procedure is corrected before proceeding. Following a procedure verbatim does not guarantee error-free or incidents-free performance, but as we learned in an earlier chapter, procedures can contain concealed defects also known as latent organization weaknesses. Therefore, comprehending the overall purpose and strategy of the procedure results in a safer and error-free outcome.

Most companies have different types of procedures depending on the purpose and use of the procedure. Some procedures are more administrative in nature; examples include human resources, procurement, etc., which provide general information for the worker. Others are required to be followed every time the activity is completed, therefore not relying on the memory of the worker. To determine which type of procedure is necessary, the procedure owner or Manager should evaluate the risk and complexity of the task, and the work group's or worker's experience and expertise with the task. Irrespective of the type of the procedure, the worker should follow the procedure or document carefully to attain the expected outcomes and results. Expectation for using each type of procedure is discussed in the next section.

Use-Every-Time Procedures are for complex or sporadic tasks for which an incorrect step or action could have direct, possibly permanent impact on safety, production, or reliability of the equipment or system. For example, Bob who works at the Ruby Rabbit Company is tasked with cleaning a tank used to make both rabbit pellets and rabbit treats. If the tank is not properly cleaned when switching batches, it will foul the next batch, causing the entire batch to be discarded and costing thousands of dollars and weeks of downtime to clean the tank; a costly error for the company. As you recall from the previous chapter, Bob is an experienced operator at the Ruby Rabbit Company. The following is the procedure Bob is to follow to start a new batch of rabbit treats[2]:

1. Verify Tank 123 has been cleaned after making pellets before starting to make treats

Chapter four: Human error defenses 73

2. Validate Tank 123 is empty by verifying that no liquid is visible in the sight-glass on the tank, and the level controller in the control room shows no liquid in the tank.
3. Verify the drain valve in Tank 123 is in the closed position.
4. Ensure the tank of Oats Blend (OB) has more than 2200 gallons, using the sight-glass and the level controller in the control room.
5. Ensure that the apple-juice tank is full by verifying the level using the sight-glass on the tank and verifying the level controller in the control room shows the tank is full.
6. Open valve 23 on Tank 123 to start the flow of materials into the tank.
7. Open valve 7 on the Oat Blend (OB) Tank, and start Pump A.
8. Verify that the level in Tank 123 is increasing by looking at the sight-glass and the level controller in the control room.
9. When the tank reaches 75% full, shut off Pump A, to stop the flow of oats, and close valve 7.
10. Verify the tank level is at 75% full in Tank 123 by using the sight-glass on the tank and the level controller in the control room.
11. Close valve 23 on Tank 123.
12. Open valve 26 on Tank 123.
13. Open valve 2B on the apple-juice tank and start Pump H to start the flow of apple juice.
14. Verify that the level in Tank 123 is increasing by looking at the sight-glass and the level controller in the control room.
15. When the tank reaches 95% full, shut off Pump H, to stop the flow of apple juice, and close valve 2B.
16. Verify the tank level is 95% full in Tank 123 by using the sight-glass on the tank and the level controller in the control room.
17. Close valve 26 on Tank 123.
18. Start the agitator on Tank 123.

As you can see, it's a fairly simply procedure to start the process of making rabbit treats, however, it is very important to the Ruby Rabbit Company that this procedure be used every time to make the rabbit treats to ensure the quality of the product. It is important to have this procedure (Use-Every-Time) followed as written. The expectations for using this type of procedure are:

- Read and comprehend each step before executing that step.
- Execute each step before starting the next step.
- Complete the steps as written in the sequence stated.
- Retain the procedure with the worker at all times while performing the steps. This prevents the worker from having to memorize the steps of the procedure.

Administrative-Use Procedures are for activities, typically administrative in nature, frequently performed, that do not require direct contact with equipment, and have no direct consequences if incorrectly executed. These procedures are within the comprehension and skills of the worker. The expectations for using these procedures are:

- Task can be executed from memory.
- Reread the procedure before performing the task, if not done recently.
- Procedure is accessible to review as necessary.

Procedure quality is vital to safety and reliability of the facility. If procedures are not accurate, complete, consistent, and usable (easy to understand and follow) it will impact the ability of the worker to properly execute them. As we have discussed before, procedures cannot be followed blindly; workers must be alert to the impact their actions might have on equipment or the system before taking any action[2].

4.2.6.1 When to utilize the tool[2]
- When operating, changing, monitoring, or investigating equipment
- When procedures exist for a task
- When the task must be done correctly the first time
- When it is unrealistic for a worker to remember every step of a procedure in order as required

4.2.6.2 At-risk practices to consider preventing[2]
- Believing a procedure is accurate as written
- Not reading a procedure before executing a task
- Initiating a procedure without ascertaining the initial conditions
- Executing a procedure action without understanding its intent
- Not providing feedback on the usability of the procedure
- Executing a procedure without identifying the critical actions
- Utilizing a checklist, attachment, or data sheet in place of the procedure
- Utilizing a procedure for an activity that the worker is not qualified to perform
- Deeming a procedure is not necessary because the worker is one of your best
- Utilizing several procedures at the same time
- Omitting steps or parts of a "routine" procedure, because those steps have not been needed in the past
- Executing a procedure knowing it is not accurate as written

4.2.7 Effective communication[2]

The objective of effective communications is the joint understanding between two or more workers, especially when communicating technical information related to facility operations or worker safety. Effective communication is likely one of the most important defenses in the prevention of errors and incidents. The oral exchange of information, especially face-to-face, is the most common form of communication – an important team attribute. Oral communications have a greater risk of being misinterpreted compared to written forms of communication. Misinterpretations are most likely to happen when the workers involved have diverse understandings, or different mental pictures of the current work condition, or use words that are unclear. Validation of information exchanged between workers must be verified to ensure the proper actions are taken. Let's use the procedure example outlined in Section 4.2.6, Procedure Use and Compliance, when Bob was asked to start a new batch of rabbit treats. Several of the steps (2, 4, 5, 8, 10, 14, and 16) required verification of the tank levels in the control room. If the communications between Bob and the control room worker were not clear, or were misinterpreted, it could have caused a problem or, even worse, an incident. How difficult is it to sometimes distinguish between some letter when spoken, for example, B and D? How easy would it be for a worker to misinterpret a direction and manipulate the wrong piece of equipment? It can and does happen. How can we help workers ensure better communications between each other? We will discuss two tools (three-way communications and using the phonetic alphabet) that will enhance better communications between workers that are easily implemented and can significantly improve communications between workers to prevent costly errors[2].

4.2.7.1 *Three-way communications*[2]

Communication between workers can occur in many ways, including face-to-face, telephone, or radio. These communications are vital to most operations because it typically takes multiple workers, working together, to operate a facility. Most operations are spread out so that the workers require some means to communicate with each other, but even if the communications are face-to face, the communications can still be misinterpreted and result in an operational error. Three-way communications can help reduce misinterpretation and reduce errors.

Three-way communication is initiated by the worker beginning the communication, as the sender of communication, who is accountable for confirming that the receiver understands the message as meant. The receiver makes sure they clearly understand what the sender is saying, then repeats back the message to the sender to ensure the message was correct. For example, Bob is in the facility starting up the rabbit-treat operations and John is in the control room. Bob and John use a radio to communicate between them. Step 2 of the procedure states "Validate Tank 123 is empty by verifying that no liquid is visible in the sight-glass on the tank, and the level controller in the control room shows no liquid in the tank." Bob is the sender and starts the communication by saying, "John, this is Bob, I need for you to verify that Tank 123 level is zero or empty on the level controller." John is the receiver. John's response is, "Bob, this is John, I understand that you want me to verify that the level in Tank 123 is zero or empty on the level controller in the control room." Bob's response is, "John, this is Bob, that is correct, I want you to verify that the level in Tank 123 is zero or empty on the level controller in the control room." What if John did not understand what Bob was asking? Then John must ask Bob to repeat the message, or add clarifying information and then repeat the steps again to ensure closed-loop communications.

By utilizing three-way communications, the chance for error can be significantly reduced. The challenge is ensuring that the third part of the communication occurs; this is typically the weakest link, because the sender may believe that the receiver heard the message correctly the first time.

4.2.7.2 Phonetic alphabet[2]

Another tool to improve oral communications is to utilize the phonetic alphabet. When working in noisy or stressful work environments it can be difficult to clearly understand multiple letters in the English language because they may sound similar. Instead of just using a letter when communicating, a word is used in its place. Utilizing a word for each letter reduces the probability that a worker listening will confuse the letters. The difficulties in clearly understanding oral communications associated with noisy areas, weak telephone and/or radio signals, and a worker's accent, are reduced by using the phonetic alphabet. Table 4.1 lists example words for each letter[2].

Table 4.1 Phonetic Alphabet

Letter	Word	Letter	Word
A	Alpha	N	November
B	Bravo	O	Oscar
C	Charlie	P	Papa
D	Delta	Q	Quebec
E	Echo	R	Romeo
F	Foxtrot	S	Sierra
G	Golf	T	Tango
H	Hotel	U	Uniform
I	India	V	Victor
J	Juliet	W	Whiskey
K	Kilo	X	X-ray
L	Lima	Y	Yankee
M	Mike	Z	Zulu

Chapter four: Human error defenses 77

Another good practice is to state what an acronym means versus just saying the acronym. Let's look at another example using the procedure noted in Section 4.2.6, Procedure Use and Compliance, step number 7, which states "Open the valve 7 on the Oat Blend (OB) Tank, and start Pump A." Again, we will listen in on Bob and John's conversation about starting the batch of rabbit treats, where Bob is the sender and John is the receiver. Bob started the oral communications using a radio. "John, this is Bob, I will open valve number 7 on the Oat Blend Tank, and start pump Alpha." "Bob, this is John, I understand you will open valve number 7 on the Oat Blend Tank and start pump Alpha." "John, this is Bob, that is correct, I will open valve number 7 on the Oat Blend Tanks and start pump Alpha." Notice that instead of using the acronym OB in the conversation both Bob and John used the term Oat Blend. Again, not using acronyms will reduce the chance for errors.

Using both three-way communications and the phonetic alphabet will significantly reduce the chance of error caused by unclear communications between workers. It also can add formality to the operation.

4.2.7.2.1 When to utilize the tool[2] Consider using three-way communications and the phonetic alphabet in oral conversations when:

- Changing the configuration of equipment or verifying the status of equipment.
- Relaying important data for work processes or equipment.
- Completing specific actions or steps using a procedure.
- Communicating alphanumeric information related to equipment.
- The sender or receiver might be misinterpreted due to sound-alike systems (B or D), high noise areas, utilizing radio or telephone communications with poor connection/reception.

4.2.7.2.2 At-risk practices to consider preventing[2]
- Sender not getting the attention of the receiver before relaying the message. Using the receiver's name helps
- Sender not trying to make eye-contact if possible, or talking behind the receiver
- Sender not being accountable for ensuring what was said was heard
- Sender not being considerate and trying to talk over someone else or attempting to talk with another worker already in another conversation
- Sender providing several actions in one message or giving too much data
- Sender believing the message was clear and not asking for verification from the receiver
- Receiver needing more input to better understand the message but is unwilling to ask for amplification of the message

- Receiver believing the communication is done and starts to takes action
- Receiver trying to multi-task
- Believing the task will get completed faster without using three-way communication or the phonetic alphabet
- Making up phonetic words instead of using the designated ones[2]

4.2.8 Place-keeping[2]

Place-keeping is when a worker actually marks steps in a finalized procedure or other document similar to what someone would do when reading a book. Place-keeping is typically utilized with use-every-time procedures, discussed in Section 4.2.6. A worker's attention continually changes from the procedure to the equipment, to indicators, to other workers, to radio/telephone traffic, and so on. If a worker is disrupted or detained, marking the last step completed provides a means to alert the worker which step is next, once they are able to restart the procedure. This is also an effective tool when the task requires more than one shift to complete; the next user will be clear where the previous shift stopped the procedure, and where the next shift needs to start. Place keeping is one tool that can help reduce the probability of an error[2].

4.2.8.1 When to utilize the tool[2]
- When use-every-time procedures are required
- When the task requires multiple shifts to complete
- Knowing at the start of a task that interruptions will occur

4.2.8.2 At-risk practices to consider preventing[2]
- Assuming it is alright to omit steps of a routine procedure because those steps have not been needed before
- Believing it is alright to somewhat follow a procedure because of experience or past success when it is a use-every-time procedure
- Noting the step of the procedure has been accomplished before actually completing the step, or noting several steps have been completed at the same time
- After being interrupted, not confirming which step of a procedure was actually completed before restarting the procedure

4.2.9 Confirming assumptions[2]

Assumptions are essential when conducting scientific and engineering work so that the issue can be refined as additional information or data is gathered or developed. The goal of a scientist or engineer is to prove that the assumptions made were appropriate and to provide objective

Chapter four: Human error defenses 79

evidence that support these assumptions. Looking back at Chapter 3 when knowledge workers were discussed, the knowledge worker must be careful not to believe an assumption is fact or forget that they made the assumption. Recalling an assumption is a mental shortcut, which can be enticing during stressful, uneasy, or unfamiliar conditions when they occur. During knowledge-based work conditions, assumptions may occur because they ease mental work by reducing the detail required. Not having the required knowledge tends to lead to incorrect assumptions that may cause errors and defects. Engineers and scientists tend to make assumptions to improve efficiency or to simply make progress with the task, until additional information is available. Be mindful of statements such as "I believe …," "We've always done it this way," "We have never had a problem with this equipment," "I'm pretty sure …," or "I think …"; these are watchwords or phrases that workers use when they are making assumptions. If assumptions cannot be verified, call a subject-matter expert to provide additional expertise to help validate inputs, assumptions, and resolve the problem. Table 4.2 provides a summary of the HPI tools and what is the appropriate setting to utilize the tool[2].

4.2.9.1 *When to utilize the tool*[2]
- In process/product evaluation meetings
- Prior to distribution of the new product
- During confirmation of document
- During calculations
- During the initial phase of the design or test
- During procurement of new equipment
- Before using initial vendor data

4.2.9.2 *At-Risk practices to consider preventing*[2]
- Utilizing assumptions and not recording them or following up on closure of invalidated assumption
- Believing assumptions are factual
- Not knowing an assumption has been made
- Having inconsistent data in two or more project/facility documents
- Having multiple sources of conflicting information and not reconciling
- Trusting past successes to defend current assumptions

4.3 *Human-error prevention techniques for teams*[2]

Often employees are working as part of a team, and, as we learned earlier, teams can also experience human errors. You cannot assume just because two or more people are conducting a task together that it will be

Table 4.2 List of HPI Tools and Applicable Setting[2]

Location	In the Plant				In the Office	
HPI Tool	Prior to Start/Restart	Perform Work	Complete Work	Prior to Start/Restart	Problem-Solving	Task Verification
Task Preview	✓					
Job-Site Review	✓					
Questioning Attitude	✓	✓	✓	✓	✓	✓
Pause When Unsure or Uncertain	✓	✓	✓	✓	✓	✓
Self-Checking	✓	✓	✓	✓	✓	✓
Procedure Use and Compliance	✓	✓	✓	✓	✓	✓
Three-Way Communication		✓	✓			
Phonetic Alphabet		✓	✓			
Place-Keeping		✓	✓			
Confirming Assumptions				✓	✓	✓

Chapter four: Human error defenses 81

correctly done. Deficiencies in performance can be caused by the interaction between team members. In a group setting, workers may not be giving their full attention to a task or action to be performed because of potential distractions from coworkers. These distractions can increase the probability of error in some circumstances. A team error is a failure of one or more workers on the team that allows another worker to make an error. This team error can happen either by the team not accepting accountability for each other, or by failure to recognize the actual capability of each team member. Another example of a team failure is when a Supervisor assumes a worker is competent and does not check the work.

Several elements can influence the interactions among workers on a team. Workers are not typically held accountable for a team's performance; therefore, some workers on a team may not actively contribute. Some workers may not get involved, assuming that they will not have to answer for their errors or any team error; these workers are sometimes referred to as the "mooch" of the team. One or more of the following social circumstances can motivate team errors.

- Halo Effect[2] – A team has ultimate faith in the competence of a worker because of their experience or education. Consequently, other workers tend not to peer-check them or even ask any questions about the actions they plan to take. This challenge is widespread in hospitals, where team members often fail to speak up when procedures or policies are not being followed. Each year it is estimated that 195,000 deaths are attributable to medical errors in hospitals, making this the third leading cause of death in the United States[1]. This does not account for many of the other errors that occur, such as administering the incorrect medication, piercing an organ during surgery causing excessive bleeding, or amputations of the incorrect limbs
- Pilot/Co-Pilot[2] – A worker does not speak up because they are not more senior or more experienced than the other worker or not in charge. The co-pilot is not willing to challenge the opinions, decisions, or actions of the pilot because of the worker's position on the team or in the organization
- Passenger Mode[2] – The inclination to "just be along for the ride" without actively analyzing the intent and actions of the worker(s) doing the activities or taking the initiative. One worker takes the action to perform the task, while the passenger worker takes an onlooker role in the activity

A good example of the Passenger Mode Error was provided in Chapter 3. Recall that Bob and his coworkers were working night shift to make rabbit treats at the Ruby Rabbit Company. Bob completed his task and went back to the control room to wait for the pressure to increase on the press machines so he could start making the treats. Bill and Robin were starting up the steam plant to heat up the ovens. Bill and Robin reported back to the control room after completing their task. Robin commented twice to Bob that is was taking a long time for the ovens to heat up, wondering if something was wrong. Bob commented that it was a cool night and sometimes it took a little longer to warm up the ovens. An explosion was heard in the plant, both Robin and Bill looked to Bob for guidance. But, as you recall, Bob was in passenger mode, appeared to be "along for the ride," not really questioning why it was taking so long to heat up the ovens or going to investigate. The three workers did not try to problem solve as a team, exchanging ideas, or really even communicating with each other.

- Groupthink[2] – Most team members work well together, they are typically unified, trustworthy, and loyal to each other. Nevertheless, at times, these traits can impact the quality of team decisions. Team members can be unwilling to communicate inconsistent data about an issue so as not to "rock the boat" and keep the "work family" happy. This can be harmful to critical thinking and problem-solving. Having team members in passenger mode, halo effect, or if one worker is very opinionated and the other members of the team choose to not "rock the boat," can have a negative impact on performing work. If these conditions exist within a group, the critical information within the team may remain concealed from other team members. Groupthink can also occur if subordinates only convey "good news" or minimize bad news to their bosses or higher-level Managers to avoid upsetting them. Some possible indications of groupthink include:
    - Delusion of imperviousness. Fosters extreme confidence and promotes unfounded risk-taking.
    - Shared justification. Ignore potential warnings signs that could cause the team to reassess assumptions before confirming previous determinations.
    - Undisputed ethics. Workers tend to disregard the ethical or moral outcomes of decisions because of the trust among team members and the belief of the team's ethics.
    - Peer pressure. Discourages workers who disagree or argue with any of the group's thoughts, reasoning, or commitments because this is not the expected behavior of loyal team members.
- Distributing responsibility can promote risk-taking among team members, when making decisions and solving problems. In some situations, a team may be more willing to take a chance as a group than each member might individually because the responsibility is dispersed among the team members. This is also referred to as

"safety in numbers" – we cannot all be wrong. This situation can drive two or more workers who think they know a better way to get something done, to take more risk and ignore procedure or policy.

The following method can reduce the likelihood for team errors:

- Preserve independent thinking among team members
- Challenge choices and actions of workers to uncover basic assumptions
- Explain to workers the challenges and pitfall of team errors, their causes, and intervention approaches
- Conduct team-improvement training
- Foster questioning attitude/situational awareness as an integral function of each task and training course
- Select a devil's advocate for problem-solving situations
- Take "timeouts" to reinforce a questioning attitude and confirm that team members understand the task and facility conditions
- Conduct a thorough task preview before starting the pre-job briefing

The challenge is to keep everyone's head in the game. Ensuring the less-experienced workers don't just rely on the more experienced worker, maintaining a questioning attitude, avoiding group thinking, and being cognitive of surroundings will help in keeping the team focused on the work to be accomplished. The next section highlights team tools to help reduce human error.

4.3.1 Peer-checking[2]

In Section 4.2.5, we discussed self-checking; before a worker takes an action, they check to ensure the proper action in being taken. A good example of self-checking is to make sure the correct valve is being opened, prior to opening the valve. Peer-checking is a similar action, only the second check is conducted by a peer that is knowledgeable of the process or procedure to prevent an error by the worker performing the action. The two workers are together at the same time and same place, before and during a specific action, to prevent an error by the worker performing the action. Both workers will verify together that the action is the correct action to perform on the correct component. Why do a peer-check? The peer may see hazards that the worker performing the action does not see and can alert the other worker. It also provides two workers looking at the same action, focused on the situational awareness, and confirming the action to be taken is the right one. Workers can request and utilize peer-checks at any time for any work situation.

4.3.1.1 When to utilize the tool[2]

Many work activities or tasks could benefit from the utilization of peer-checking, such as the following:

- Verify that a critical step is completed
- Verifying an action that could either be irreversible or create unwanted outcomes is correct
- Verifying test data with the acceptance criteria
- Verifying that the correct major component is either operating or shut off
- Ensuring the correct piece of equipment is returned to or removed from operation
- Ensuring the correct parts or correct components are used before maintenance
- Ensuring similar equipment that could be interchanged or installed incorrectly is properly installed

4.3.1.2 At-risk practices to consider preventing[2]

- Having a peer that is not familiar or experienced with the task perform the check
- When the peer is not watching the performer or believes the performer will not make a mistake
- When the peer cannot see the component
- Having a peer that is considerably less experienced than the performer and may be very reluctant to correct the performer
- When the performer takes an action, before the peer is ready to complete the peer-check
- When the performer is less engaged in completing the action, assuming the peer will identify any problems

4.3.2 Independent verification[2]

Independent verification provides confirmation on the status of a piece of equipment (i.e., valve ABC is closed), the correct document is being utilized, or calculations are accurate for safe operation. Independent verification is performed by two knowledgeable workers, separated by time and distance from the action being taken. For example, one worker changes the position of a valve by themselves (no one is present at the time) and another worker verifies the position is correct (without the presence of the first worker). Independent verification is utilized when taking an action to change a piece of equipment, (like the position of a valve), avoiding using the wrong document, or an error in a calculation could subsequently cause an adverse impact to the facility if the improper condition remained undetected.

The independent verification process typically has a higher chance of identifying an error than peer-checking because the verifier is not involved in modifying the component's state or developing the document and their knowledge of the system, component, or work condition is unchanged by the performer. The verifier actually checks the status of the piece of equipment or document without relying on the other worker.

Independence occurs because the verifier is not biased by the thoughts or actions of the performer. Dividing the actions of the performer and verifier by time and distance encourages freedom of thought. The performer, while setting up the requested state, does not communicate with the verifier, and the verifier cannot either see or hear what the performer is doing.

4.3.2.1 When to utilize the tool[2]
- During positioning of important equipment related to safety
- After maintenance of equipment prior to restart
- When modification of equipment could negatively impact plant safety
- Validation of design and safety related calculations
- Verification of safety related operating or maintenance procedures

4.3.2.2 At-risk practices to consider preventing[2]
- Both workers will be involved in the independent verification, traveling to the location of the piece of equipment together, before the initial action.
- Both the performer and verifier are coworkers and/or co-located on the same task.
- The performer is less engaged in the task, assuming the verifier will catch any errors.
- The verifier only checks what the procedure stated to do and did not look for other indications that the status was correct, for example, verified that the valve appeared to be closed, but did not verify no flow in the line.
- The verifier is less experienced than the performer and is unwilling to question the correctness of the action, design, calculation, etc.

4.3.3 Flagging[2]

An incident can occur if a worker incorrectly operates or performs maintenance on the wrong piece of equipment. This can occur if a worker starts an activity and selects the wrong piece of equipment, because several similar-looking pieces of equipment are in the same location or close to the same location. Additionally, events have been known to occur when the worker takes a break or was distracted, and resumes work on an adjacent, similar – but wrong – piece of equipment. If multiple components are physically located near other similar-looking components, flagging or marking the component can help the worker consistently identify the correct component. Utilizing self-checking, the worker clearly marks the correct piece of equipment with a flagging notation, which aids the performer visually to return to the correct piece of equipment during the activity or after a distraction or interruption. Figure 4.1 provides an example of flagging similar pieces of equipment that are not being worked on[2].

Managers and Supervisors should have a list of approved methods for flagging. Examples of methods could include use of colored adhesive dots, ribbons, colored tags, rope, or magnetic placards. Flagging methods should not interfere with facility equipment, including the observation of operating indicators, i.e., meters[2].

4.3.3.1 When to utilize the tool[2]
- When handling an equipment near similar-looking components
- While working on equipment that will be operated several times
- When the need for flagging is identified during task planning

4.3.3.2 At-risk practices to consider preventing[2]
- Utilizing similar flag methods for pieces for equipment to handle and not to handle
- Before utilizing flagging methods, not self-checking or peer-checking
- Utilizing flagging methods that block any plant indicators or interfere with equipment
- Using a flagging method that can be moved easily or become displaced, such as a post-it-note
- Forgetting to remove a flagging method after completing the task

Component A

Component B

Component C

Figure 4.1 Example of flagging. (Courtesy of PRESENTERMEDIA.)

4.3.4 Turnover briefing[2]

A turnover briefing is the relaying of vital plant information, activities, and responsibilities between workers or crews. A turnover briefing provides the oncoming worker(s) an accurate status of the plant and any additional pertinent information (open excavations, visitors in the plant, etc.) before assuming responsibilities and starting work. A good turnover briefing enables each worker to understand where things stand at the beginning of their shift and what workplace conditions are expected during the shift. Turnover briefings may also be appropriate if major activities are occurring and workers switch positions for lunch breaks, or for maintenance tasks that take multiple shifts to complete or involve multiple work groups.

The purpose of any turnover brief is to convey information critical to the successful continuation of a project or activity passed from one group or worker to another in a manner that reduces interruption of work activities and encourages safe and efficient work completion. Turnover briefings can differ in detail and format depending on the risk of the task and the nature of the work being conducted.

Turnover briefings should be brief and simple but thorough and accurate. Turnover briefings could involve visually looking at plant indicators, reviewing procedures in process, walking-down equipment or activities in the plant, and discussing the ongoing activities during the upcoming shift. Workers discuss the work conditions. The oncoming worker or group is given a chance to ask questions and resolve concerns. The more effective turnover briefs are written to ensure the oncoming worker clearly understands the status of the plant and important information is documented for future reference. A Lean tool that is frequently utilized is a checklist. A checklist will guide the workers through a comprehensive review of plant status, procedures in progress, etc. – items that are important to the safety of the facility. Turnover briefings facilitate communications between the oncoming worker and off-going worker such that the oncoming worker is fully capable of assuming the duties and responsibilities and planned tasks, before handing over responsibility to the oncoming worker[2].

4.3.4.1 When to utilize the tool[2]
- Prior to shift change
- When responsibilities are shifted between workers, work groups, or organizations (handoffs)
- When conducting new, critical, or complex activities that cross several shifts
- When transferring responsibilities for activities in progress

4.3.4.2 At-risk practices to consider preventing
- Attending a turnover briefing while attempting to conduct another task or taking a phone call, or any activity requiring attention
- Not conducting a face-to-face verbal briefing
- Omitting critical information or the explanation for decisions made
- Not recording activities and/or other important information
- Allowing distractions during the turnover briefing that disrupt the workers
- Shifting accountability to an oncoming worker who is not prepared to assume the responsibilities
- Rushing the turnover briefings, and not allocating sufficient communications in the shift schedule
- Not utilizing a checklist

4.3.5 Post-job review[2]

After an activity is completed, a good practice is to conduct a post-job review. During the review it's important to determine what went well, what did not go as well as planned, and identify opportunities for improvement the next time the activity is performed. Comments on the initiation, planning, execution, and actual work conducted is important for the Management Team to understand. Areas to explore during a post-job review include:

- Procedure compliance
- The work process (how well was the activity planned, any work-arounds required, etc.)
- Equipment utilization
- Tool and/or supply problems
- Minor human errors that require Management's attention

Such conditions could become a latent organization weakness if not corrected. When workers discuss the information, Managers have the opportunity to reduce weaknesses with processes, programs, procedures, and job-site situations that could prevent event-free performance.

A post-job review typically involves a face-to-face meeting between the worker(s) and the Supervisor(s), giving the workers the opportunity to provide feedback on how the work was performed. An effective post-job review discusses lessons learned to improve the activity if performed in the future and supports completion of the paperwork related to the job.

Typical topics discussed during post-job reviews include the following:

- Outcomes that were unexpected or a surprise
- Procedures, and/or other work documents usability and quality

Chapter four: Human error defenses

- Worker knowledge and skill deficiencies
- Insignificant errors that occurred during the activity
- Unexpected job-site conditions or workarounds
- Supervisory support effectiveness
- Suitability of tools and other resources
- Work preparation/planning and scheduling quality
- Identification of important lessons learned to document for future activities
- Other issues that arose during the work activities

To underpin the effectiveness of post-job reviews, Supervisors should provide timely feedback to workers on the resolution of significant issues identified during reviews[2].

4.3.5.1 When to utilize the tool[2]
- After completing any activity where difficulties occurred
- At the completion of a non-routine or important work task
- After finishing a new task for the first time
- After completing routine work where improvements were incorporated

4.3.5.2 At-risk practices to consider preventing[2]
- Not conducting a post-job review after working on important facility equipment or a highly complex task
- Principal workers not available for the post-job review
- Not allocating time for the post-job review or the review conducted in a rush
- No methodology to follow-up on issues identified during the post-job reviews
- No follow-up with principal worker's for significant issues
- Not conducting fact-to-face post-job reviews
- Significant issues not being addressed or documented for future pre-job briefings

4.3.6 Project planning[2]

Project management principles provide a framework for working on tasks with distinct start and end dates. Typically directed by administrative instructions, project management tools support keeping the scope, objectives, and deliverables aligned with the facility strategic plan. Additionally, a disciplined and organized approach reduces rework, assumptions, and oversights by encouraging designers to carefully plan and scope the work. Likewise, project management supports reduce stress and associated time pressures by fostering communication, foresight, and planning activities.

Project management involves activities (or tasks) associated with attaining a set of project goals while enhancing the use of limited resources (time, money, space, people, etc.). These actions could include initiation, planning, controlling, and closure. Utilizing a systematic approach to the management of any project activities will reduce the potential for human error.

An effective project management approach will help maintain a reduced risk of failure over the lifetime of the project. The primary risk of failure arises from the uncertainty at all phases of a project. The potential for human error is one aspect of uncertainty during a project completion. A worker cannot consistently and constantly maintain awareness and remember all the intricate details of a project, especially complex projects. Providing a method for robust planning, monitoring, and control of a project, supports achieving project objectives on time, within budget, and without defect[2].

4.3.6.1 When to utilize the tool[2]
- While design work is ongoing
- For work not otherwise directed by administrative procedures
- Work activities involving extra personnel (typically subcontractors)

4.3.6.2 At-risk practices to consider preventing[2]
- Limited or infrequent communications
- Not getting stakeholders' or customers' input during the planning process and not seeking their commitment/ownership for the project
- Not controlling the project plan, allowing scope creep and unapproved changes
- Not addressing human performance controls during planning phase
- Not documenting a project plan
- Not documenting decisions in real time
- Not addressing resource issues for competing priorities
- Not clearly defining or documenting project scope
- Frequently switching project team members

4.3.7 Problem-solving[2]

Problem-solving is a knowledge-based performance condition. There is no obvious answer, no familiar example, rule, or skill to recall. Problem-solving entails connecting cause-and-effect relationships in reverse order to decide why the problem happens. Problem-solving necessitates methodical examination to recognize potential causes. Problem-solving techniques that work best are organized, easy, and notable, and match data-gathering techniques and analysis tools. The fishbone method, described in Chapter 2, Section 2.5.1.3, is one example of a defined problem-solving technique.

Chapter four: Human error defenses

If workers experience a situation that is unfamiliar, the preference is to pause the activity, have a questioning attitude, and involve another worker or Supervisor with the problem. Given that the worker is in knowledge-based performance situation, the chance for an error is particularly high (1 in 2 to 1 in 10). Getting others involved can support effective problem-solving and provide additional insight.

A fishbone diagram, also known as a cause-and-effect diagram is a visualization tool for classifying the potential causes of a problem in order to distinguish its root causes. There are four main steps to developing a fishbone diagram:

1. Identify the problem
2. Identify the major factors involved
3. Identify the possible causes
4. Analyze the diagram

4.3.7.1 When to utilize the tool[2]

- When troubleshooting
- Throughout conceptual design
- Conducting causal analysis
- In knowledge-based performance conditions for workers
- When conditions cannot be determined using questioning attitude

4.3.7.2 At-risk practices to consider preventing[2]

- Describing a problem in terms of potential reasons
- Trying a resolution or changes before describing the problem adequately
- Assuming problems can constantly be resolved with a procedure change and/or training
- Waiting for long-term corrective actions, instead of implementing short-term actions to prevent reoccurrence
- Defining a cause, if no root cause can be identified
- Employing a corrective action when uncertain of the cause
- Ruling out potential causes without justification or facts
- Permitting a worker to overshadow a problem-solving method with their suggestions
- Evading disagreement

4.3.8 Decision-making[2]

Decision-making is a progression approach used to predict the possible effects of a decision. Workers review all of the possible alternatives and potential impacts and then select the best option. A worker who follows a systematic decision-making approach can safeguard against rule-based and knowledge-based errors.

Conservative decisions are focused on ensuring the facility is in a safe state and not to ensure that the near-term production goals of the organization are met. Typically, this determination is clear. Nevertheless, for truly knowledge-based conditions, this may not be obvious. A thoughtful, systematic method endorses better decision-making. Decisions should:

- Explain the goal
- Identify alternatives
- Include suitable examination of those alternatives in achieving the goal
- Foster a plan to execute the identify resolutions
- Identify methods to quantify the efficiency of the plan

Decision-making can happen from both a short-term and a long-term perspective. Sometimes, depending on the conditions, decisions must be made quickly, while some decisions may be delayed to ensure adequate time for a formal analysis. Irrespective of the time limitations, facility and worker safety necessitate conservative decision-making. The following methods can enhance conservatism:

- Utilize all accessible data, challenging the urge to overlook inconsistent or non-conforming information
- Utilize workers who can provide a supplemental understanding, such as including Management and front-line workers in decision-making
- Decrease uncertainty as much as possible, with a focus on facts, and confront assumptions
- Sustain safety notwithstanding production stresses
- Contemplate the accumulative risk (significance) of the decision. Create contingency activities

Leaders can appoint a "devil's advocate"[2] function to enhance conservative decision-making by the team, and to oversee and challenge the team's decision-making process. A devil's advocate maintains an attentive observation for potential faults and omissions because most workers focus on achieving the activities or tasks[2].

4.3.8.1 When to utilize the tool[2]
- When an error could have significant negative outcomes
- Throughout the initial or conceptual design segment of a critical action
- While creating project work plans

- Throughout product review meetings
- While conducting troubleshooting actions
- When organizing product evaluation presentations
- Throughout an engineering assessment and operability evaluation
- Throughout the final segments of a root-cause analysis
- When acquiring unlike substitute components because like components are not accessible

4.3.8.2 At-risk practices to consider preventing[2]
- Determining decisions before outlining the goal
- Utilizing biased, unorganized techniques, such as perception, knowledge, and brainstorming, for risk-important challenges
- Executing a decision without truly grasping the risks
- Not taking swift corrective actions to avoid recurrence of a condition before executing a longer-term corrective action

4.3.9 Contractor oversight[2]

Most companies utilize vendors/contractors to support facility operations for limited work scopes or fixed-time special projects. Contractors can be at a risk of contributing to significant incidents at a facility if not trained or accustomed to the work culture. Most companies offer general employee training, but the training is typically not robust enough to offset the contractor's lack of facility experience, specifically related to industrial safety, industrial hygiene, and human performance.

Contractors require the same coaching and mentoring as facility workers when they are working with their workforce. Contractors must comprehend that their work practices, especially human performance, must meet the same requirements as those expected of facility workers. Remember, these contractors are working side-by-side with your work force, and if an event occurs, it can impact the contract workers, as well as your worker and facility equipment[2].

4.3.9.1 When to utilize the tool[2]
- When a prescribed procedure for directing interactions with contractor workers does not exist
- While planning for contractor services
- When buying new equipment
- While acquiring contractor services
- After the contract has been awarded, but before the work actually starts
- Throughout the contractor's implementation of the work scope
- When sending equipment back to the contractor for overhauling, restoration, troubleshooting, or maintenance

- In advance of concluding the contracted activity
- After conclusion of a task
- If there is indication, or doubt, of inadequate completion of work or outcomes
- Prior to utilizing contractor provided data

4.3.9.2 At-risk practices to consider preventing[2]
- Presumption that the contractor is the "expert" and will not make errors
- Presumption that the contractor workers have the same standards as your facility workers
- Inadequately confirming or testing contractor-provided designs
- Conducting inadequate oversight of contractor work practices
- Presuming the contractor understands the impact of changes to their standard product

4.4 Human error prevention techniques for managers[2]

The techniques in this section can be to be utilized by Managers and Supervisors to support detection of latent organization weaknesses. These are primarily unnoticed defects in organizational processes or values that generate a workplace situation that enable errors (error precursors) or decays the reliability of defenses (flawed defenses). Unnoticed organizational weaknesses can aggravate human performance. Latent errors or situations are often challenging to identify. After being generated, typically they appear to fade away, but instead accumulate in the system. Because of their concealed trait, restraining the time these weaknesses exist is challenging. Managers should actively identify and correct weaknesses with defenses promptly. Several tools that can be used to assist management will be discussed in this section[2].

4.4.1 Benchmarking the competition and industry leaders[2]

Benchmarking is an activity of comparing the performance of one's own organization for a specific topic or function against another organizations that performs better in the same area, and learning what the other organization does to achieve exceptional performance. This comparative analysis could include the identification of positive practices, performance standards, and pioneering thinking or methodologies. Benchmarking is an activity of gauging products, services, and practices against robust competitors or those companies established as industry leaders.

Chapter four: Human error defenses

For benchmarking to be successful, Managers must assess the various characteristics of their own organization's performance to ascertain the performance. With these outcomes known, identify external organizations whose performance is excellent and who are inclined to share performance information. The best way to learn from another organization most often necessitates extensive communication between organizations and a trip to the host organization to observe how business is conducted. With the knowledge gained from visiting the host company, an improvement plan can be developed on how to adopt the best performance practices discovered.

Benchmarking is an influential management tool because it conquers the "Paradigm Blindness." "Paradigm Blindness" is the manner of thinking: "Of course, this is the best way to do it. It's the way we have always done it"; or "We are so different, we cannot learn from other organizations, because they do not understand our situation."

Benchmarking allows the organization to see a different approach, new methods, ideas, and tools to increase efficiency. It provides a platform for workers to see, first-hand, another approach that is working and is effective in achieving results. It also demonstrates that there is another way to accomplish work[2].

4.4.1.1 When to utilize the tool[2]

- If the current process is unproductive, unsuccessful, too costly, or too risky
- When errors, near-misses, and mishaps increase in a specific zone
- When a gap has been identified between actual performance and Peak Performance in a given area
- If the Management Team identifies the desire to significantly improve how business is done

4.4.1.2 At-risk practices to consider preventing[2]

- Failure to identify processes or practices that require refinement
- Failure to take the time necessary to collect the required data and rushing into demonstrating improvement phase
- Failure to identify "best in class" benchmarking companies, just ones that conduct business differently or marginally
- Failure to recognize major distinctions in management systems or the processes of the two organizations that can impede successful implementation of best practices
- Reluctance to communicate information with another company
- Failure to develop and implement an improvement plan after completing benchmarking with the other company

4.4.2 Observation of work and behaviors[2]

Observing work in the plant or field is a good technique to obtain valuable information for how work is actually performed, as well as how the organizations support work at the job site. The purpose of observing behaviors and workflow is to evaluate the quality and effectiveness of work planning, work practices, and work performance. The purpose is not to condemn or to judge workers. The real intent of these observations is to pinpoint opportunities to optimize the overall flow of work (work environment, tools, etc.), while observing workers while they are performing work.

The observations should be from the start of the activity to completion of the activity. Focus not only on the worker's practices, but observe the job-site environment, look for potential hazards and the controls implemented to mitigate those hazards, and the overall flow of the work activity. After completion of the observation, document observations for trending purposes to support identification of strengths and weaknesses over time. Observations of worker behaviors can identify organizational weaknesses that may not be apparent by any other methods. When the Management Team spends time in the facility with the personnel while performing work, coaching and mentoring the workers, performance improves and error rates tend to decrease[2].

4.4.2.1 When to utilize the tool[2]
- Validate if the organization supports individuals' performance at the job site
- Emphasize preferred behaviors and coach worker to improve skills
- Note strengths and weaknesses of work activities to increase performance
- Detect and record visible latent organizational weaknesses

4.4.2.2 At-risk practices to consider preventing[2]
- Not being prepared for the observation, not understanding the activity being observed, not knowing what to observe or when to conduct an observation
- Lacking the skills to be critical or being disrupted during the observation
- Asking the easy questions, not evaluating for actual understanding
- Allowing poor practices to continue or not pausing when at-risk behaviors are observed
- Limiting observation to worker's behaviors, not looking at the job-site situations, administrative procedures, and practices that support worker performance
- Conducting an observation and not providing feedback to the workers and Supervisor on what was observed, losing an opportunity for coach and mentor

- Not documenting what was observed (any findings) or not trending performance over time to determine if the crew is improving
- Not correcting behaviors, job-site conditions, administrative weaknesses, or error-likely situations after being identified

4.4.3 Self-assessments[2]

Self-assessment is a review of the practices and performance of one's facility/operation. Assessments can be formal or informal, but provide the opportunity to identify potential improvements by contrasting current practices and results with preferred practices, results, and standards. Self-assessments can provide a means of detecting latent weaknesses in the organization and operations. The best individual to conduct an assessment is someone who understands how things really work in the organization. Managers should encourage workers to conduct self-assessments to enable continuous improvements and detection of drift in compliance with procedures and policies. An assessment can also help identify situations that may adversely impact defenses or provoke error-likely situations that set workers up for failure.

Self-assessments should cover all aspects of your operations, from administrative activities to maintenance. This will also support your "Lean Lens" reviews presented in Chapter 2. Most companies develop an annual schedule that supports assessment planning. To support developing the annual schedule, review the events that occurred over the past 18 months: if there are any common procedures or policies, conduct a self-assessment on those procedures or polices. Review injuries for the past 18 months, look for commonalities; again, assess those common attributes. Also consider a periodic timeframe to review all procedures and practices, (i.e., three to five years); this will assist in identifying any potential for drift from desired practices[2].

4.4.3.1 When to utilize the tool[2]
- If trending indicates a potential increase in human performance errors
- If noted that current practices and results are skewed with preferred goals and objectives
- To determine if the operation is complying with safety codes, regulations, and requirements

4.4.3.2 At-risk practices to consider preventing[2]
- Selecting areas to assess that are known to be easy to access, i.e., administrative areas only
- Not identifying credible objectives for designated areas
- Not determining consistent and accurate measures for the designated area

- Not evaluating all applicable work areas during a predetermined timeframe
- Having assessments completed by a worker that is either not familiar with the work or is not given adequate time to complete the assessment
- Not being critical of the operating/process assessed and unwilling to identify findings or provide feedback to the work group
- Not developing or implementing corrective actions after an assessment identified areas that required change or improvements

4.4.4 *Performance indicators and application*[2]

Performance must be measured, typically utilizing indicators or metrics, to ensure organizational expectations are met. The reason for these measures is to gauge whether production goals, safety, and other critical activities are occurring in a manner that supports outstanding performance. If indicators are outside of a preferred range and/or performance, this might imply a minor issue, giving ample time to correct before a major event occur.

There are two types of indicators used by most industries:

- Lagging[2] – Measures will not predict future performance or accomplishments, but are a gauge for results or consequences that indicate what has been accomplished and where you are with respect to company goals and objectives
- Leading[2] – Measures that look at organizational "health,", which can forecast performance and accomplishments; measures may also address system conditions

The process of choosing suitable indicators that are meaningful and useful to track performance is challenging and requires prudent review, continuous enhancement, collaborations, and comprehension. Performance indicators are routinely selected because of ease to obtain the data, i.e., easy to count; however, most of the time these indicators provide limited awareness as to how the organization is currently performing and provide no insight for future performance. Some examples of leading indicators include: industrial-safety number of employee observations per month, percentage of overtime per month, absenteeism per month, backlogs (e.g., procedure revisions, failed equipment repairs, work orders, maintenance items), number of management observations in the plant per month, or number of Lean improvement suggestions submitted per employee per month.

A lagging measure is monitoring results, either good or bad, which may be a meaningful indicator for your operation, but it does not

Chapter four: Human error defenses 99

demonstrate if current or past performance is sustainable. Lagging indicators can help identify seasonal challenges, for example, the 90 days of summer. Several examples of lagging indicators include: industrial safety lost-time injuries per 200,000 man-hours worked, number of vehicle accidents per miles driven, rework (typically defined as any activity that must be redone or repeated because of poor quality or it is incorrect that results in loss of time, labor, money, or other resources within a particular period of time), recurring corrective actions, or number of accidental needle-sticks per total number of needle-sticks.

Trending methods or practices critical to success serve as very useful leading indicators because they are a prediction of things to come. The next step after creating performance indicators and collecting the data is to trend the data to identify slight drifts or shifts in performance before a major event occurs. Many industries note an increase in injuries during the summer months (June to August), which could be caused by distractions, heat, worker absent for vacations, etc. Not only are selecting the correct indicators imperative, but analyzing the data to really understand the performance of the organizations is needed to provide the real benefit and value from the indicators[2].

4.4.4.1 *When to utilize the tool*[2]
- When deciding which indicators are needed to evaluate overall performance
- To record error trends, incidents, and near-misses over a timeframe

4.4.4.2 *At-risk practices to consider preventing*[2]
- Selecting parameters that are easy to obtain or measure
- Utilizing a range of performance indicators that have not been approved by the Management Team
- Collecting and maintaining indicators that are not analyzed or trended
- Trying to maintain too many indicators or not having quality indicators

4.4.5 *Independent oversight*[2]

Appraisals of facility activities by impartial organizations or agencies offer a chance to uncover "blind spots"[2] for facility management and personnel that might otherwise continue to be concealed. Independent Quality Assurance departments, consultants, and regulatory assessments offer a chance to find these latent situations. Independent oversight activities can identify situations, processes, and practices that do not match the best practices of industry standards and identify conditions that can lead to declining facility performance if not corrected.

Most industries conduct benchmarking exercises, looking for the best in the business and then comparing how their performance and operations stack up. An external independent assessment is basically the opposite of a benchmarking exercise. The goals of either benchmarking or independent assessment are to learn how to improve practices and processes to continue to strive for Peak Performance. The company must embrace the assessment and recognize that the external team is present to help challenge the company to achieve even greater performance.

Often independent assessments are requested by Senior Leadership not Production Managers. Therefore, the conclusion is that if we need to be accessed by an "outside" organization, something must be terribly wrong. Most Managers and Supervisors have confidence in their organization and the work performed. Senior Leadership of a company may find it challenging communicating to Managers or Supervisors the importance of having someone else look at the operations, ask the difficult questions, and play the "devil's advocate. Managers and Supervisors tend to become very defensive and overprotective of their work and staff. It's a natural human tendency to question an outsider evaluation. It can be very challenging to hear an outsider tell you that your operation is below standard, your workers are not knowledgeable, procedures and policies are inadequate, etc.; however, it may be the best intervention for the team to take a self-critical look, and really be honest about the status of the operations and personnel.

Individuals with a wide range of experiences and knowledge that are outside the organization can identify possible weaknesses and trouble spots easier than the workers within the organization. The leadership can take advantage of the observation and potential issues from an independent review[2].

4.4.5.1 When to utilize the tool[2]
- To make potential improvements in the overall operation and maintenance of the facility that are more in line with best industry practices and high standards
- To reinforce the importance of standard organizational processes and company values
- Subsequent to identifying the need to improve the business or operation
- At the request of the leadership to ascertain issues within an organization

4.4.5.2 At-risk practices to consider preventing[2]
- Missing the opportunity to learn and improve performance because the leadership team has become protective and unwilling to listen and learn from an outside review, therefore, limiting the learning and potential enhancements that could result from the findings and overall observations

- Limiting the attendance at the out-brief of the independent review, not including the workers in reviewing the findings and observations, and not allowing input on the corrective actions. These actions limit the ownership of the workforce and commitment to make changes and improvements
- Assuming the independent review team did not understand your facility and operations and therefore is only doing a superficial analysis of the findings that could prevent the real identification of systemic or programmatic weaknesses, resulting in no real changes
- Overlooking the need to conduct self-assessments or management reviews to verify the that corrective actions are being implemented and really internalized

4.4.6 Work product evaluation[2]

These evaluations foster face-to-face communications between Supervisors and subject-matter experts, procedure writers, and other knowledgeable workers. These interactions help not only the developers of the materials to improve, but provide an opportunity for others to learn and strengthen the overall team. It also provides an excellent opportunity for Supervisors to coach and reinforce expectations in a positive manner. These reviews can also provide insight into potential weaknesses of the organization, such as training. To support continuous improvement, Managers can provide a checklist for workers to use to ensure the work products meet standards. Another tool to support continuous improvement is grading of the work products and then trending[2].

4.4.6.1 When to utilize the tool[2]
- Routinely by the leadership team (Managers, Supervisors, etc.)
- If a negative trend in work products is noted
- As required by policies/procedures
- While conducting apparent cause evaluations and root-cause analyses

4.4.6.2 At-risk practices to consider preventing[2]
- Not performing self-critical reviews of work products because you "trust" your team to generate good products
- Performing superficial reviews of work products only to achieve a "quota"
- Not reinforcing the expectation to conduct these reviews by the entire team, and then not holding the leadership team accountable to ensure completion
- Conducting the review late in the cycle, causing excessive rework or forcing late delivery of the product because of the rework.

4.4.7 Change management[2]

When significant changes are being planned it is a good idea to establish a change management process. A change management process is a systematic approach or planning process to outline the direction of change, align workers and resources, and implement the designated change throughout the organization, whether large or small.

The systematic approach must involve the Leadership Team to ensure that the change is effectively implemented and completed in a timely manner. Day-to-day activities are not typically subjected to a change management process, just sizable organizational changes that impact a large number of workers or involve a significant reorganization. Utilizing effective change management processes can help the leadership identify unintended consequences and impacts to policies or procedures, reduces the potential for errors caused by changing the way operations/business is conducted, and limits the overall impact to workers[2].

4.4.7.1 When to utilize the tool[2]
- To lead the change process and implementation of improvement initiatives
- When significant organizational changes are proposed

4.4.7.2 At-risk practices to consider preventing[2]
- Not articulating a clear vision and expectation for change
- Not engaging workers that will be impacted by the change
- Not sharing the need for the change and helping workers understanding the desired outcome
- Not taking the appropriate time to understand the new values, attitudes, and beliefs needed to affect the change in a positive manner
- Failing to communicate that a change will be made, the purpose for the change, and the overall value to the organization
- Failing to carry out the change management plan, being in a hurry, and not allowing time for the workers to adjust to the change

4.4.8 Reporting errors and near misses[2]

Organizations that practice human performance principles and concepts request and expect workers to report errors and near misses. The leadership team utilizes the feedback to pinpoint organizational challenges and assist workers and teams to learn from their mistakes to enable improved performance in the future. Recording failures and trends provides the organization with the opportunity to gain lessons learned without the negative impact of having an accident. It is important to have an established process-and-error reporting system. The organizational culture

Chapter four: Human error defenses

must be one where workers feel free and supported, and where Managers foster and appreciate the reporting of errors and near-misses[2].

4.4.8.1 When to utilize the tool[2]

- When seeking to gain data on the extent of active errors occurring and the nature of these errors
- Looking for ways to change practices that prevent errors from occurring
- To reinforce training and development actions to enhance performance
- When the leadership team is ready to take the facility operations to the next performance level

4.4.8.2 At-risk practices to consider preventing[2]

- Implementing an error reporting system without first instilling a learning culture, i.e., encouraging reporting of error and near misses without fear of discipline
- Not addressing workers who exhibit a fear of reporting, clearly communicating the reason the leadership team is interested in the reporting of error, and near-misses to improve the overall performance
- Obtaining the data and not implementing improvement actions
- Utilizing the reported data in a manner not intended and breaking the trust of the workers involved
- Not providing timely feedback on errors reported to the workers and the proposed improvement actions

4.5 Critical steps[2]

A critical step is defined as a procedural step or sequence of steps or an action that, if completed inappropriately or incorrectly, will cause irrevocable damage to plant equipment, or people, or that will significantly impact facility operation. The definition includes a work-step/activity that if incorrectly completed has an instantaneous adverse consequence that cannot be negated or reversed.

Examples of critical steps include:

- Amputating a limb
- Clicking "accept" for terms and conditions on purchasing a house
- Removing a fuse in an electrical cabinet
- Opening a valve on a process line
- Intravenously injecting medication
- Entering a confined space
- Igniting fireworks

The goal of any operation is to ensure positive control at critical steps when error-free performance is vital. Utilizing human performance tools does not assure Peak Performance, but workers can significantly reduce their odds of erring by using the tools. Several tools can be employed to reduce errors and maintain positive controls at critical steps. Some of the tools previously discussed are shown in the next section, with examples of how the tool can be used for critical steps, including: task preview, pre-job briefing, and procedure use[2].

4.5.1   Task preview[2]

Task preview provides the opportunity to define the critical steps in the task and discuss what tool can be used to reduce the chance of committing an error. A good tool to utilize is S-A-F-E-R[2]. This process helps workers systematically identify and take positive actions to reduce the risk of an error.

- S – summarize the critical action/steps
- A – anticipate potential errors at each critical step and applicable error precursors
- F – foresee likely and worst-case outcomes if an error were to occur at a critical step
- E – evaluate methods to control or possibilities for each critical step to avert, identify, and restore operations from errors and diminish the significances
- R – review prior occurrences to lessons learned related to the particular activity

4.5.2   Pre-job briefing[2]

A pre-job briefing is a gathering of workers and Supervisors prior to performing a job to discuss the tasks, critical steps, hazards, and related safety measures. This gathering helps workers to better understand the task(s) to be completed, the potential hazard(s), and the critical step(s), i.e., what the actions are that *must* be completed correctly to prevent a significant event from occurring. Workers explain the task's objectives, roles, and responsibilities, tools to be utilized at each critical step to ensure error-free performance, and resources necessary to complete the activity. Clearly understanding the task required to be performed, the roles and responsibilities of each worker, the hazards, and the measures that will be taken to reduce the hazards will certainly help improve the error-recognition potential. During the briefing, other important information to present includes precautions, limitations of the system, controls that will

Chapter four: Human error defenses   105

be employed, contingency plans, and relevant lessons learned from previous jobs or from other industries.

A good pre-job briefing increases everyone's knowledge of critical activities and gives workers time to mentally practice performance of critical steps. Pre-job briefings help workers prevent unanticipated situations in the facility and emphasize the concept that there are no "routine" activities and that every task, no matter how many times performed, can be slightly different or something can go wrong[2].

4.5.3   Procedure use[2]

Procedures prescribe the steps or actions needed to complete a task. If procedures are used as intended it will reduce the potential for errors and events. However, experience has shown that an event can still occur even when workers use procedures, when important steps are skipped, or omitted. When performing an activity where multiple steps are required to be completed, some steps, i.e., critical steps, are more important to the overall success than others. As an example, let's look at the steps required for making a cup of coffee.

(1) Pour 8 ounces of water into the coffee-maker
(2) Place the paper filter in the holder
(3) Place 4 scoops of ground coffee in filter and place in coffee-maker
(4) Place coffee-pot on burner of coffee-maker
(5) Plug cord of coffee-maker into 110-volts wall outlet
(6) Turn the switch on to start the coffee-maker

This is an easy procedure for making coffee. What are the consequences of skipping any of these steps? No cup of coffee to drink when it is expected. What is the critical step in this procedure? It is step number 6 – before that, more water or ground coffee could be added, or one could stop and just not make a cup of coffee. But once the switch is turned to the "on" position, the coffee-maker is on and making coffee. Let's explore examples of how the critical step could have been denoted in the procedure and highlighted to the worker along with human performance improvement tools to support error-free performance.

(1) Pour 8 ounces of water into the coffee-maker
(2) Place the paper filter in the holder
(3) Place 4 scoops of ground coffee in filter and place in coffee-maker
(4) Place coffee-pot on burner on coffee-maker
(5) Plug cord of coffee maker into 110-volts wall outlet
(6) **Critical Step** Turn the switch to start the coffee-maker ** **Critical Step** ** Use peer-checking to ensure error-free performance

Another example:

> (1) Pour 8 ounces of water into the coffee-maker
> (2) Place the paper filter in the holder
> (3) Place 4 scoops of ground coffee in filter and place in coffee-maker
> (4) Place coffee-pot on burner on coffee-maker
> (5) Plug cord of coffee-maker into 110-volts wall outlet
> [Critical Step] (6) Turn the switch to start the coffee-maker [Use peer verification]

The goal is to ensure that the worker recognizes the step is a critical step and is prepared to implement a human performance improvement tool to reduce the likelihood of an error. These tools should be covered during the pre-job briefing as well as who will be responsible for performing the actions associated with the human performance tools. It is also a good practice to clearly mark the steps off when completed[2].

4.5.4 How to get started using critical steps[2]

The logical place to start utilizing critical steps is when the scope of the task is being developed. Start by asking a couple of questions:

- What is the worst thing that could happen?
- What can be done to prevent the worst thing?

Understanding the hazards and controls will help identify critical steps in the work process. The next step is to identify the controls that can be used, as well as the critical steps, and human performance tools that can be employed. This is also a good place to review previous lessons learned from other experiences.

Let's look at another example. The Ruby Rabbit Company is going to make a new flavor of rabbit treats by starting up an old line that has been out of service for almost a year. The task will be to clean out the old line and conduct maintenance on the equipment to ensure it is operable, and restart the line making a new-flavor rabbit treat. The consideration of potential hazards includes: equipment could have failed with material still in the system that has degraded over time that could leak out or spray onto the workers when the system is opened; the material for the new recipe may not be compatible with the previous materials; and corrosion of the equipment could lead to catastrophic failures (causing a component or structure to fail). To develop the scope of work for the task, walk down the production line looking for obvious issues – leaking components, structural issues, exposed wiring, etc. – making a note of each issue. Next,

Chapter four: Human error defenses 107

review diagrams and drawings of the process system to identify power sources and other input and outputs for the line. Perform a walk-down to verify all hazards have been identified. For example, a motor was missing on one of the tanks; this motor needs to be replaced so the line can operate properly. Another tank appears to contain liquid; the oven's thermocouple wires appear to be worn, etc.

Once the hazards have been identified, a control must be developed to mitigate each hazard. For example, the motor missing on a tank, or a lock-out is needed to de-energize the line so a new motor can be installed. The liquid in the tank needs to be sampled before being removed to ensure proper disposal and understanding what personal protective equipment will be needed to perform the task, the thermocouple wiring needs replacing, another lock-out is needed.

Review the procedure for this task and identify the critical steps involved and the human performance improvement tools to reduce the potential of errors during the task preview to ensure everyone understands the task to be completed, roles and responsibilities, and expectations.

(1) ** Critical Step ** Install a lock-out for Tank 51 to replace the motor ** Critical Step **
(2) Lift the motor into place. The motor weighs 90 pounds and will require a team lift
(3) Connect the wiring for the new motor
(4) Remove lock-out for motor on Tank 51
(5) ** Critical Step ** Install a lock-out to replace the thermocouple on Oven A ** Critical Step **
(6) Replace thermocouple wiring for Oven A
(7) Remove lock-out for Oven A
(8) Sample contents of Tank 57 using goggles, face-shield, gloves, and apron. Collect 50 milliliters of material for analysis
(9) Remove all debris from the line in preparation for restart of the line
(10) Once sample results have been received on Tank 57, drain the content into three 55-gallon drums
(11) After filling the drums, move drums to the proper storage location depending on the analysis
(12) ** Critical Step ** Energize the entire system to start completing functional checks of each piece of equipment ** Critical Step **

The goal is to understand at what point a step could cause irrevocable damage to the plant equipment or people and significantly impact facility operation, or to identify a step that if completed incorrectly has an instantaneous adverse consequence that cannot be negated or reversed. Preventing errors in this scenario should focus on the errors that workers make while in the knowledge-based performance mode as described in

Chapter 3, Section 3.5.3, because the task that the team is doing is unfamiliar. During work execution, the human performance objective is to foresee, prevent, or find active errors, especially at critical steps, where error-free performance is vital. By understanding the work scope, identifying critical steps (where the actions taken must be error free), and applying human performance improvement tools the potential for errors and events is significantly reduced.

References

1. HealthGrades quality study: Patient safety in American hospitals. 2004. http://www.providersedge.com/ehdocs/ehr_articles/Patient_Safety_in_American_Hospitals-2004.pdf
2. Department of Energy (DOE) (2009) *Standard Human Performance Improvement Handbook*, Volume 2: Human Performance Tools for Individuals, Work Team, and Management. Department of Energy, Washington DC.

chapter five

Operations excellence

5.1 Introduction

What is Operations Excellence? If you work for a university you might define Operations Excellence as the method by which you manage and operate all facets of the institution, including admissions, facility maintenance, and waste management. If you work in a restaurant, Operations Excellence may be reflected in the quality of the food and facility cleanliness. If you work for the military, Operations Excellence could mean the discipline, respect, and rigor needed to successfully achieve a mission. Depending upon the company or institution, Operations Excellence can be defined in many ways; however, what is fundamental to all definitions is that Operations Excellence consists of programs and processes necessary to establish a disciplined and structured operation that ensures mission success. The Operations Excellence process identifies and defines mission-critical operations, activities, desired behaviors, and attributes needed to achieve performance beyond routine operations.

Operations Excellence is not defined by each employee, or by select individuals within the organization, but rather by a collaborative team of people, which may include representatives from Senior Management and Leadership, First-Line Supervisors, and representatives from the workforce. The importance of including employees from within the organization or company in establishing the program enhances fundamental knowledge of the program itself and instills ownership during program implementation. The end objective of an Operations Excellence program is to identify those operational activities, tasks, processes, and programs that need additional focus, discipline, and rigor to ensure a consistent outcome or product. Institutions and companies that have implemented a form of an Operations Excellence Program have also exhibited a decrease in the likelihood and consequences of failures of human performance and of technical and organizational systems.

In essence, an Operations Excellence Program builds upon the streamlining of work processes and systems accomplished through implementation of a Lean process, but also adds a level of discipline to work processes and tasks that have been improved upon by a human performance improvement program. These programs are preventative, add value, and are representative of continuous improvement programs commonly used in industry today.

The benefits of implementing an Operations Excellence Program have been demonstrated in both the government and commercial sectors of the nuclear business for many years. In addition, sustained and consistent operational performance, and improvement in how processes and programs are implemented, have been successfully demonstrated through various levels in the organization (i.e., performing daily operational tasks or managing key processes within your business structure) when implementing an Operations Excellence Program. So how do you develop an Operations Excellence Program?

Within the Peak Performance Model, the traditional Operations Excellence approach has been tailored to recognize that mission-critical processes vary among companies, organizations, and institutions. As depicted in Figure 5.1, an Operations Excellence Program consists of five core functions, which identify, evaluate, and organize mission-critical operations and activities into manageable program elements and associated processes. These program elements and associated processes are identified as being mission-critical and monitored to ensure consistent operational performance and quality.

Operations Excellence Program Elements represent common themes, programs, or processes that originate from mission-critical operations and activities. Implementation of the Operations Excellence Program targets increased rigor, discipline, and focus when performing

Figure 5.1 Operations excellence process and core functions.

Chapter five: Operations excellence 111

mission-critical operations and activities. Additionally, the Operations Excellence Program presented in this book acknowledges, and recognizes, the impact organizational culture has on implementation of a successful Operations Excellence Program. If the culture of the company or organization does not promote the fundamentals of a learning and healthy safety culture, then the impact and success of implementing an Operations Excellence Program will be marginal. The negative impact may not be obvious, but over time impacts from a poor safety culture result in higher injury rates, diminished productivity, and reduced product quality. It can never be over-emphasized that the organizational and safety cultures of a company or institution underpin the ability to achieve mission success.

5.1.1 Core function one: Identify mission-critical processes

Core Function One of the Operations Excellence process consists of identifying company or organizational mission-critical processes. Mission-critical processes are those components of a business or institution that have a direct impact on whether the mission or operation will be successful. A general rule-of-thumb is that most companies have the following mission-critical processes:

- Procurement of equipment, materials, and supplies
- Product research and development
- Operations,
- Equipment and building maintenance,
- Business administration
- Management and workforce

This list can be tailored to mission-critical processes associated with a company or institution no matter the size of the entity. There are many operational and business processes associated with companies and institutions, but only those processes that could significantly impact the company or institutional mission should be identified. In addition, depending upon the company's organizational structure and products produced, those operations designed as mission-critical can also change.

As part of Core Function One, it is recommended the company or institution establish an Operations Excellence Team [may be assigned as an ancillary job assignment]. The purpose of the Operations Excellence Team is to formally define and document the company's Operations Excellence Program and to monitor and improve implementation of the program over time.

There are many approaches that can be used when establishing an Operations Excellence Team, but the best team is one that is based on a collaborative group of personnel who represent different work groups across the company (i.e., Company President, Plant Manager, Operator, Electrician, Shift Supervisor) and provide different opinions as to what are mission-critical operations. Working sessions, including possible use of a facilitator, are promoted, to encourage team-member collaboration and thought-development.

5.1.2   Core function two: Identify mission-critical operations and activities within each mission-critical process

Core Function Two of the Operations Excellence process is to identify mission-critical operations and activities (including associated documents) within each mission-critical process. Mission-critical operations and activities include those key processes and programs that establish and implement operational or activity-specific requirements. The identification of Operations Excellence mission-critical operations and activities is a necessary activity and can be enlightening to both the Management and workforce if personnel are active participants in the process and Senior Management sponsors and encourages organizational learning. There is no defined number of mission-critical operations and activities that must be identified, but the list should be encompassing enough to ensure all regulatory, environmental, safety and health, and Corporate processes are addressed. Several examples of mission-critical operational or activity work processes that have policies, procedures, or guidelines associated with specific work activities are:

- A company or organization's Project Execution or Strategic Plan. This may include specific goals and objectives flowed-down through a Corporate entity. These documents define how a company or organization will operate to achieve mission success.
- Business administration policies and procedures associated with human resource functions. Company and Corporate policies related to safety and health procedures, procedure compliance, employee rights, and employee discipline.
- Environmental compliance, which includes all company and Corporate policies of operational activities. Operational activities may include items such as:
  - Routine surveillances to support environmental permit requirements.
  - Operational requirements associated with regulatory requirements.
  - Communications within the plant and external stakeholders
  - Emergency notification and protocol.

Chapter five: Operations excellence 113

Once mission-critical operations and activities have been identified, along with their associated procedures, then the Operations Excellence Program Elements can be identified, defined, and monitored.

5.1.3 Core function three: Establish Operations Excellence Program Elements

Core Function Three of the Operations Excellence process is the identification and establishment of Operations Excellence Program Elements. A product of implementing Core Function Three is development of a cross-walk that identifies what operational and work activities are critical to mission success, along with associated policies. The established Operations Excellence Program Element should be associated with more than one mission-critical operation and/or activity.

For example, an effective communications program is critical to the success of a company or organization's mission; however, communications are implemented in a number of ways and at various levels within the company or organization. The company or organization may have policies and procedures that include communications as a primary component of emergency notification. General communication guidelines are often defined for public-address systems or protocol to be used when responding to media inquiries. The Operations Excellence Team should brainstorm and use cognitive decision-making in defining program elements that take into account common themes, processes, or programs used across multiple mission-critical processes. Listed below in sections 5.1.3.1 to 5.1.3.4 are several examples of identified Operations Excellence Program Elements.

5.1.3.1 Organization and administration

The organization and administration of a company, and, in particular, operations and maintenance organizations, are almost always considered as mission-critical. The administration of the organization establishes policies, programs, and procedures that define an effective operations organization including:

- Roles, Responsibilities, Authorities, and Accountability (R2A2) of Management and Personnel, including those associated with operations and maintenance (tailored to the size of organization or company)
- Operational programs and procedures that promote Management and employee ownership and accountability in the safety program and safe performance of work
- Qualification requirements defined for positions of the organization or company

5.1.3.2 Communications

How Management and workers communicate within the company is also almost always considered a mission-critical operational activity. Most companies desire communication to be straightforward, clear, and accurate. Chapter 4, Section 4.2.7 provided examples of effective communications techniques. Several examples of how effective communication is considered a mission-critical activity are:

- Identification of routine and non-routine communication systems and mechanisms (in today's work environment that would include both verbal and electronic notification and communication systems)
- Configuration and administrative control of communications mechanisms, including the use of automated, electronic, and public-address systems
- Desired notification processes both internally, Corporate, and with outside stakeholders (e.g., local fire department)
- The use of acronyms and terminology consistent with executing the operational mission

5.1.3.3 Training programs

The Operations Excellence Team should review their training programs and identify what training programs are critical to achieving mission success. Those operations or activities identified as being critical will vary depending upon the mission of the company or organization. In addition, by using a collaborative team to identify mission-critical training, the company or organization will gain insight as to which processes the workers identify as vital to success versus Management, which may identify different training programs they deem as vital to mission success. Several examples of the different types of training employed by companies and institutions are:

- Training of personnel on operational equipment
- Qualification training versus general employee training
- Leadership training
- Management training
- Regulatory-driven training (e.g., Hazardous Waste Operations and Emergency Response [HAZWOPER])
- Continuous education training
- Company or Corporate business and workplace training

5.1.3.4 Equipment configuration management status and control

One of the most important processes of mission-critical systems is the ability of a company to ensure consistent work practices associated with equipment lineups and configuration, and ability of the operations to

Chapter five: Operations excellence

adapt to potential operational changes. This is extremely critical because everyday issues arise during the performance of work by people, and operational changes or improvements. The ability of a company or organization to be able to adapt to these changes directly contributes to mission success. Several examples of activities that address status of equipment and operational process control are:

- Proper operational equipment alignment, and a defined process for implementing preventative and corrective equipment maintenance
- Configuration control, approval, and awareness of hazardous energy equipment lockouts and tagouts and other hazardous operations, through equipment testing, startup, operation, and shutdown
- Operational safety and engineered limits that ensure a consistent end product and compliance with all company and regulatory requirements
- Control of temporary equipment modifications and temporary systems
- Configuration control and distribution of engineering documents

These are just a few examples of commonly identified mission-critical operations or activities. Table 5.1 is an example crosswalk, which depicts several identified mission-critical processes, mission-critical operations and work activities, and associated identified Operations Excellence Program Elements to be monitored and evaluated. Table 5.1 is not all-inclusive; the crosswalk represents an example of how to define Operations Excellence Program Elements.

Once the crosswalk is developed, then each program element is further defined by identifying the implementing company or Corporate policies or procedures that are key to successful implementation of the Operations Excellence Program Element, along with desired process or behavioral attributes that are to be incorporated into implementation of the operation or work activity. Table 5.2 presents an example program element that has been further defined by desired process or behavioral outcome. Often these behaviors are derived from the company mission and core values. Company core values can be integrated into the program elements by ensuring their concepts and definitions are embedded in how mission-critical policies and procedures are implemented.

The policies and procedures associated with each Operations Excellence Program Element are actively monitored through Core Function Five to ensure the necessary rigor and discipline is applied to achieve mission success. Alignment of the company-specific policies and procedures against the Operations Excellence Program Elements enables the company or organization to increase focus on processes that are critical to mission and company success.

Table 5.1 Example Operations Excellence Crosswalk

Mission-Critical Processes	Mission-Critical Operations and Activities	Operations Excellence Program Elements
Procurement of Equipment, Materials, and Supplies	• Procurement of Materials and Supplies • Quality Control of Materials and Supplies • Product Inventory • Renting Equipment • Purchasing Equipment	• Company Procedures • Quality Control • Product and Equipment Configuration Management • Communications and Notifications • Company Training
Product Research and Development	• Product Material Content and/or Flowsheet(s) • Product Quality Control • Research and Development • Product Recalls	• Product and Equipment Configuration Management • Company Procedures • Quality Control
Business Administration	• Environment, Safety, and Health • Company Strategic Plan • Company and Corporate Business Policies and Procedures • Company Business and Accounting Services • Human Resources – including Organizational Charts • Employee Safety Council • Workforce and Management Training • Safeguards and Security	• Organization and Administration • Communications and Notifications • Training • Company Procedures

Table 5.2 Example Operations Excellence Program Element: Organization and Administration

Operations Excellence Program Element: Organization and Administration Functional Element Owner: Plant or Operations Manager	
Desired Process or Behavioral Attribute	Company Policy and/or Procedure
Roles, Responsibilities, Authorities, and Accountability (R2A2) of the organization or company are clearly defined within policies and procedures	• Specific Corporate or company policy that requires the establishment of defined R2A2s • Company-specific project execution plan • Company procedure that requires R2A2s defined within each company policy and procedure • Work procedures • Human Resource position descriptions
Ownership and accountability of the safety program and the safe performance of work is demonstrated through company policies and procedures	• Company safety policy • Employee-led safety committee charter • Employee Rights and Responsibilities Policy • Company mission and vision statement
Personnel critical to completion of the mission are clearly identified and managed throughout the organization or company, including development and succession plans of key positions	• Company-specific project execution plan • Executive management plan that includes the identification of key positions critical to mission success (these positions may or may not be in management) • Human Resource position descriptions • Company Training Plan
Training programs have been established and effectively implemented throughout the organization and company	• Company-specific project execution plan • Company-specific training program document • Defined position descriptions with required skills clearly identified and mapped to required training programs and classes • Special skill requirements are identified and associated training developed

5.1.4 Core function four: Implement the Operations Excellence Program

Core Function Four in the Operations Excellence process is implementation and sustainment of an Operations Excellence Program. Core Functions One through Three were focused on development of a program; Core Function Four is focused on program implementation and takes into consideration training and communications that may be needed when launching and implementing a new company or organizational program.

Most people do not realize the amount of work needed when launching a new company program. Program documentation of an Operations Excellence Program should include, at a minimum, the following:

- Purpose and focus of the program with Senior Management sponsorship
- Inclusion of the Operations Excellence Charter and R2A2s associated with implementing and sustaining an Operations Excellence program
- Definition of mission-critical operations and activities
- Identification of Operations Excellence Program Elements and key company policies, procedures, and documents that are credited with fulfilling desired attributes and requirements
- Defined process for evaluating and identifying improvements to the program

All of these items should be included and addressed in a company-specific Operations Excellence Program document.

Many companies develop a communications plan to target communications to different work groups because of the difference in R2A2s among the various job positions. For example, communication of the program to Management may include emphasizing their R2A2s associated with implementing company procedures. A typical communication plan should contain the following:

- Identified target audience
- Key message(s) to be communicated and understood
- Examples of how the communication relates to the worker's job, including examples of how operational risks and value added to worker's jobs are related
- How the communications would be transmitted (e.g., weekly publication, daily pre-job, during task review, one-time all employee message)

It is recommended the Operations Excellence Team work with communications personnel to understand how to effectively communicate to the different skill levels within the company. In addition, risk communications should also be integrated into the communication messaging, such that employees and Management will understand how an Operations Excellence Program enhances and improves workplace safety and productivity. Ensuring effective and targeted communications of the program will prepare employees for program implementation; however, development of a training program for employees is also recommended to assist Management in implementing the program within their organizations.

Development of a training plan will define the knowledge and performance expectations of employees when rolling out an Operations

Excellence Program. The training plan does not need to be elaborate, but should address:

- Clear definition of what the Operations Excellence Program means to Company X.
- Benefits to employees, and the company, when implementing an Operations Excellence Program.
- Clear definition of the relationship between mission-critical processes, mission-critical operations and activities, and program elements. The training should identify those processes, procedures, attributes, and behaviors that are to be monitored for consistent and improved performance.
- How the job categories relate to and support Company X Operations Excellence Program Elements.

It is recommended that a basic training class be developed as part of the rollout of the program. Continuous training should be incorporated, and highlighted, into everyday work activities and should be considered to be integrated into the onboarding process when hiring employees.

5.1.5 *Core function five: Evaluate and implement improvements*

Core Function Five of the Operations Excellence process includes evaluation of how well the program is being managed, implemented, and whether it continues to add value to the company or success of the organizational mission. Benefits of evaluating the program include:

- Confirmation that identified program elements are still critical to the mission
- Ensured consistency in the performance of tasks and their resultant outcome
- Heightened awareness and attention to detail when performing work tasks
- Improvements observed in the efficiency of work performance
- Reduced operational costs
- Continuous positive customer feedback on product quality or service

A collaborative team should be used in performing the evaluation and recommendations should target improvement in different work groups (i.e., employee, First-Line Supervisors, Mid-Level and Senior Management) and different organizations. A company may choose to have the Operations Excellence Team perform the evaluations, or they may prefer independent personnel to evaluate, using "a fresh set of eyes."

There are a number of tools that can be used to assess adequacy of an Operations Excellence Program. Listed are a few examples of several types currently used when evaluating program performance. Many of these tools were discussed in greater depth in previous chapters (i.e., Chapter 2, Chapter 4).

- Documentation reviews and confirmation that expected performance, including expected behaviors and practices associated with the company policies and procedures, are still relevant and consistent with the company mission and performance expectations.
- Formal assessments with defined criteria and lines of inquiry are performed by personnel within and/or outside the company.
- Informal assessments, both worker- and Supervisor-performed, which focus on desired attributes and behaviors.
- A formal field-observation program, which uses checklists designed to evaluate implementation of procedures and policies that support an Operations Excellence Program.
- Interviews with both individuals and work groups.
- Company and workgroup surveys.

The Operations Excellence Team should develop assessment tools as part of the formal establishment of the program and identify and define types of methods to be used for evaluating program performance. In particular, the tools should target desired behaviors and performance expectations and can be tailored to be used by various members of the organization (i.e., Operators versus Plant Management). Having the tools, such as checklists, developed and standardized upfront, allows use of the same set of questions by different organizations within the company, and can be used to evaluate and compare performance amongst divisions.

The Operations Excellence Team, with sponsorship from Senior Management, identifies the frequency for evaluation. For example, the Operations Excellence Team determines that weekly Operations personnel will informally use a program element checklist once a week to focus on one mission-critical operation or activity and provide real-time feedback to the work crew.

The size of the company or organization may help, by default, define the frequency by which evaluation of the program is conducted, along with the methods chosen for performing the evaluation. If the company has fewer than 10 employees, then the Operations Excellence Program can be tailored to focus on those mission-critical operations; subsequently, evaluation of the program may only be annually, or conducted as part of a monthly operational inspection. Alternatively, if the company is greater than 100 employees, the program may be larger and evaluations of the program could be more frequent or more in-depth.

Throughout this book, various methods have been presented that further describe these assessment tools. An example checklist for evaluating an

Operations Excellence Program is presented in Table 5.3. The checklist was developed to support evaluation of an Organization and Administration Program Element (associated with Table 5.2). Each of the lines of inquiry was associated with, or related to, an attribute, behavior, operation, or work activity deemed as mission-critical. As always, "one size does not fit all," and processes described in this chapter have to be tailored to the individual company or institutional objectives and mission.

Once an evaluation of the program has been performed, areas for improvement should be identified and actions taken to address weaknesses. In addition, reevaluation of the program is recommended to confirm processes and procedures being targeted remain mission-critical and value added from the program is still being realized. Because of the tremendous value of establishing an Operations Excellence Program, Operations Excellence is one of the four functional elements in the Peak Performance Model.

5.2 Application of an Operations Excellence Program

The desired outcome of implementing an Operations Excellence Program is a high-performing Management team and employees who work toward achieving a common operational mission and vision. If the company operations are dependent upon human capital, then implementation of an Operations Excellence Program will result in consistent performance by personnel and improved work efficiency. As described in Section 5.1, an Operations Excellence Program is implemented through five core functions. So how is the process applied to a business or organization, and what does it look like? Development and application of the Operations Excellence Program for the Ruby Rabbit Company are demonstrated in the next section. The Operations Excellence process presented in Figure 5.1 was used in development of the program.

5.2.1 Core function one: Identify operations excellence mission-critical processes at the Ruby Rabbit Company

The Ruby Rabbit Company has been manufacturing rabbit pellets and treats for over 30 years. It has grown over the years and become a prominent supplier in the manufacturing of rabbit pellets and treats. Currently, the Ruby Rabbit Company employs almost 1,000 people, who perform various jobs such as procuring supplies, operating mixing and processing equipment, maintenance and infrastructure, administration, and Management staff. Annual operational revenue equates to almost $3,000,000 per year. Over the past three years, Management has noticed several problems, including a decrease in the level of workforce productivity and product quality, and Management appears to be reactive to operational situations as opposed to having a proactive operations mentality.

Table 5.3 Example Organization and Administration Program Element Checklist

Line of Inquiry	Yes	No	Comments
Do workers understand their roles, responsibilities, accountability, and authorities associated with performing their job?			
What are some example R2A2s that workers can explain?			
Can workers explain examples of when they recognize they are working outside their defined work scope?			
Do workers understand what actions are required when they identify procedures are not being correctly implemented?			
Do workers understand the Company's safety policy?			
Does Management understand the behavior expectations and notification requirements in case of an emergency?			
Have goals and objectives of the Company's strategic plan been integrated into the organization and administration of the company?			
Have succession plans been developed for those job positions which are critical to achieving mission success?			
Do all company processes and procedures, which implement the Organization and Administration Program Element, identify and support the desired performance outcomes?			
Do employees feel they are adequately trained, and empowered, to perform their job function?			
Do organizational charts accurately reflect R2A2s and reporting requirements?			
Do personnel understand the company's mission and how their job supports mission completion and company success?			
Do workers believe their job is being performed in the most efficient and effective manner?			

Chapter five: Operations excellence 123

Several Managers from the Ruby Rabbit Company were recently at an industry conference and heard about another company that had implemented an Operations Excellence Program that had resulted in significantly increasing profits, customer satisfaction, and product quality. Senior Management at the Ruby Rabbit Company decided to adopt this program and tailor it to their operations. Five employees from the Ruby Rabbit Company were chosen to be members of a team that would be responsible for development of the Operations Excellence Program.

Representatives from the union, administrative personnel, Senior Management, Shift Supervisor, and maintenance met and brain-stormed to identify mission-critical operations and activities and associated attributes. One of the team members was familiar with the Operations Excellence process and volunteered to write a charter for the team and to lead the team in developing the program. The result of the initial brainstorming session was the identification of mission-critical processes. If critical operations were impacted, then the mission and subsequent value of the Ruby Rabbit Company would diminish. Mission-critical processes of the Ruby Rabbit Company are summarized in Column 1 of Table 5.4, which contains an excerpt from the Ruby Rabbit Operations Excellence Program Crosswalk.

5.2.2 *Core function two: Identify critical operations and activities within each mission-critical process at the Ruby Rabbit Company*

Following identification of the Ruby Rabbit Operations Excellence mission-critical processes, the Ruby Rabbit Operations Excellence Team performed Core Function Two and identified mission-critical operations and activities required to ensure mission success. The Operations Excellence Team reviewed mission-critical operations and activities to determine which work documents and procedures were essential to ensuring that the Ruby Rabbit pellets and treats were of good quality, produced on time, with a focus on cost efficiencies. Several of the team members were surprised to discover that many documents thought to be critical to the company mission were not, and several that were identified as being critical were not clearly understood by workers.

Mission-critical business and operations work processes and procedures of the Ruby Rabbit Company are summarized in Column 2 of Table 5.4, which contains an excerpt from the completed Ruby Rabbit Operations Excellence Program Crosswalk.

Table 5.4 Ruby Rabbit Operations Excellence Program Elements and Associated Cross-Walk

Ruby Rabbit Company Mission-Critical Processes	Ruby Rabbit Company Mission-Critical Operations and Activities	Ruby Rabbit Company Operations Excellence Program Elements
Procurement of Equipment, Materials, and Supplies	• Procurement of Materials and Supplies • Quality Control of Materials and Supplies • Product Inventory • Renting Equipment • Purchasing Equipment	• Company Procedures • Quality Control • Product and Equipment Configuration Management • Communications and Notifications • Company Training
Product Research and Development	• Product Material Content and/or Flowsheet(s) • Product Quality Control • Research and Development • Product Recalls	• Product and Equipment Configuration Management • Company Procedures • Quality Control
Operations	• Operational Valving and Alignment of Chemical Tanks 1 through 22 • Operational Checklists • Daily Surveillance of Tanks 1 through 22 • Emergency Response and Notifications • Daily Pre-Job Briefings • Operations Training • Operations Communications and Use of Public Address and Other Communication Systems • Work Control System	• Operating Practices • Communications and Notifications • Training • Control of Interrelated Processes • Company Procedures • Product and Equipment Configuration Management • Quality Assurance • Organization and Administration

(*Continued*)

Chapter five: Operations excellence 125

Table 5.4 (Continued) Ruby Rabbit Operations Excellence Program Elements and Associated Cross-Walk

Ruby Rabbit Company Mission-Critical Processes	Ruby Rabbit Company Mission-Critical Operations and Activities	Ruby Rabbit Company Operations Excellence Program Elements
Equipment and Maintenance	• Operations manual for Chemical Tanks 1 through 3 • Operations manual for Pellet Mixer 1 • Operations manual for Acid Tank Wash • Bi-monthly Assessment of Pellet Mixer 1 Valves and Electrical System • Identification and Labeling of Equipment • Preventive and Corrective Maintenance • Products Specifications • Maintenance Training	• Routines and Operating Practices and Procedures • Communications and Notifications • Company Procedures • Company Training • Product and Equipment Configuration Management • Quality Assurance • Organization and Administration
Business Administration	• Invoicing and Billing for Materials and Services • Environment, Safety, and Health • Company Strategic Plan • Company and Corporate Business Policies and Procedures • Company Business and Accounting Services • Human Resources – including Organizational Charts • President's Safety Council • Management and Safety in the Field • Workforce and Management Training • Company Policies and Procedures • Network Administration and Security	• Organization and Administration • Communications and Notifications • Training • Company Procedures

5.2.3 Core three: Establish Operations Excellence Program Elements for the Ruby Rabbit Company

The Operations Excellence Team of the Ruby Rabbit Company evaluated both the mission-critical operations and policies, procedures, and work documents that establish and implement company programs. Based on that analysis, listed is an excerpt of the Ruby Rabbit Company's Operations Excellence Program Elements:

- Access control
- Company procedures
- Product configuration management
- Control of interrelated processes
- Equipment labeling
- Communications and notifications
- Shift routines and operating practices

The list is just an example of some program elements associated with the Ruby Rabbit Company. There is an endless list of possibilities as to the type and number of program elements to be identified; however, the elements should be simplistic, manageable, and make sense to Management and the workforce. Table 5.4 is an excerpt from the Ruby Rabbit Operations Excellence Crosswalk. Table 5.5 is a crosswalk for the Operations Excellence Program Element: Organization and Administration. Within the table are identified process and behavioral traits desired to emphasize and instill discipline in performance of the task.

5.2.4 Core function four: Implement the Ruby Rabbit Operations Excellence Program

After three months of development, the Ruby Rabbit Company Operations Excellence Team was ready to formally establish and implement an Operations Excellence Program. Essentially, this step of the process was "rolling out" the program to all company or organizational personnel. The formal charter for the Operations Excellence Team was formally communicated. Senior Management held an All-Hands Meeting and communicated how the program would benefit the employees and the company in improving the safety of the workplace and productivity. In addition, "The Ruby Rabbit Company Operations Excellence Program Description Document" was issued. The program description document defined how the program would be managed, implemented, and defined R2A2s for program maintenance, implementation, and improvement.

At the Ruby Rabbit Company, a formal communications plan was developed that targeted communicating how mission-critical operations

Chapter five: Operations excellence 127

and activities require additional focus and rigor in performing work tasks. Table 5.5 is an excerpt from the Ruby Rabbit Company Operations Excellence Program Crosswalk pertaining to Program Element: Organization and Administration. Desired attributes were identified for this program element and company-specific documents were identified and incorporated into the Operations Excellence Program, which supported mission-critical processes and desired performance attributes.

The communications and training plans were tailored to communicate and train personnel at various levels in the Ruby Rabbit Company on how company-specific documents and procedures, used by the Ruby Rabbit Company workforce, are relevant to their job, and why it is important to strive for minimum errors in performing their job

Table 5.5 Ruby Rabbit Operations Excellence Program Element: Organization and Administration

Ruby Rabbit Company Operations Excellence Program Element: Organization and Administration Functional Element Owner: P. Rabbit, Chief Operating Officer	
Desired Process or Behavioral Attribute	Company Policy and/or Procedure
Roles, Responsibilities, Authorities, and Accountability (R2A2) of the organization or company are clearly defined within policies and procedures	• RR-105, *Ruby Rabbit Company Mission Organizational Roles, Responsibilities, Authorities, and Accountabilities Program Document and Matrix* • RR-109, *Ruby Rabbit Company Operations Execution Plan* • RR-203, *Ruby Rabbit Company Position Descriptions* • RR-210, *Ruby Rabbit Company Employment Requirements*
Ownership and accountability of the safety program and the safe performance of work is demonstrated through company policies and procedures	• RR-501, *Ruby Rabbit Company Safety Policy* • RR-522, *Ruby Rabbit Company Employee-Led Safety Committee Charter* • RR-501, *Ruby Rabbit Company Employee Rights and Responsibilities Policy*
Personnel critical to completion of the mission are clearly identified and managed throughout the organization or company, including development and succession plans of key positions	• RR-109, *Ruby Rabbit Company Operations Execution Plan* • RR-123, *Ruby Rabbit Company Executive Management Plan* • RR-203, *Ruby Rabbit Company Position Descriptions* • RR-400, *Ruby Rabbit Company Training Plan*

5.2.5 Core function five: Evaluate and implement improvements at the Ruby Rabbit Company

After six months of implementing the Ruby Rabbit Operations Excellence Program, the Company President requested the Operations Excellence Team initiate their first evaluation of the program. The purpose of the evaluation was to be two-fold: 1) establish a performance baseline, and 2) evaluate training effectiveness and areas for improvement. The Operations Excellence Team reviewed the program document to determine the most appropriate method for performing the evaluation. The Operations Excellence Team determined that they would use checklists when observing work in the field, and also conduct focus-group interviews with workers and Supervisors over a period of 60 days. As part of initial program development, standardized tools such as checklists were developed. Results of the evaluation determined additional emphasis was needed in Operations Excellence Program Elements: Communication, Shift Routines and Practices, and Training. An Operations Excellence Improvement Plan was developed and presented to Management, to target improvement in these three program elements over the next year.

chapter six

Culture
The bedrock for sustainable organizational performance

6.1 Introduction

It is widely recognized that culture has far-reaching implications for organizations; from the way work is performed, relationships with customers, developing and implementing procedures, and integration of the diversity of workers from various backgrounds to include those from around the globe. The United States is recognized as a nation of immigrants. As a nation of immigrants, the culture of an organization can become complex and difficult to manage in an integrated system that facilitates organizational success.

Early on, many authors and theorists provided similar definitions of what they believed constituted organizational culture in broad terms. These definitions had various similarities and contributed to formalizing the way organizational culture is defined today. However, Edgar Schein is noted and recognized by many as one who has significantly contributed to defining culture and the role it plays in organizations. Schein's definition is as follows:

> a pattern of shared basic assumptions that the group learned as it solved its problems of external adaptation and internal integration that has worked well enough to be considered valid and, therefore, to be taught to new members as the correct way to perceive, think, and feel in relation to those problems.[1,3]

Many authors and theorists later defined culture in terms that are somewhat aligned with earlier definitions. The most significant outcomes of those definitions are that all of these authors agree that the culture of an organization has significant impacts on performance. In addition, the writings of these authors support the premise that culture encompasses the practices, values, beliefs, behaviors, and patterns that are evident in the organization and forms the identity of the organization. Developing a culture that can facilitate business growth is very challenging because of the diversity of people and the different rituals they bring to the organizations and varying subcultures that underlie the overall culture of an organization.

6.2 Whistleblower

In work cultures where employees are actively engaged, they trust their leadership team, and feel free to raise concerns when they arise; therefore, the concept of whistleblowing is typically not seen. Perhaps one of the most compelling reasons why some companies are focusing on the culture of their organization is due to the influx of activities surrounding potential negative exposure from presumed unethical behaviors by company leaders and employees. These negative perceptions can create company losses in many ways, to include loss of reputation and revenue due to the fines that may be imposed. When employees expose potential company wrong-doing to an external agency this act is known as whistleblowing. A whistleblower is an individual who has access to company information, and shares that information, usually with someone in authority, to legally explore or investigate the claim that is being made by the individual. There are employment laws that have been enacted to protect workers in situations where they feel the need to expose internal Corporate wrong-doing and/or unethical behavior. In the government sector, the Whistleblower Protection Act of 1989 is one such law enacted to protect federal workers when they report organizational misconduct.

Once an individual has filed a "Whistleblower" case, they are protected from retaliation by their employer. Under this act, employees can file complaints about violations such as waste and fraud, abuse of authority by leaders or the company, violation of laws, or safety to workers or the public. Penalties can be imposed on an organization for violation of the law.

Although the Act of 1989 as stated does not cover employees of companies that are not associated with or working under a government contract there are other laws that have been enacted to protect whistleblowers working for various industries. Examples of whistleblower protection laws are listed in Table 6.1.

There are also other avenues that have been taken to expose a company for wrong-doing, such as anonymous contact with the news media. Many companies have enacted policies that provide protection for employees that bring issues to the attention of management. Whistleblowers have become an integral part of keeping companies and leaders honest in their treatment of workers, caring for their health and safety, and protection of the environment.

Table 6.1 Whistleblower Laws and Regulations

The Law	Applicability
Affordable Care Act (ACA)	Applies to employees who report violations such as discrimination based on the denial of coverage due to a preexisting condition, an insurer's failure to provide appropriate rebate, or an excess premium or denial of health insurance subsidies
Clean Air Act (CAA)	Applies to reporting of violations regarding air emissions from area, stationary, and mobile sources
Comprehensive Environmental Response, Compensation and Liability Act (CERCLA)	Protects employees who report violations relating to improper cleanup of hazardous waste sites, accidents, spills, and any releases of pollutants and contaminants
Consumer Financial Protection Act (CFPA)	Applies to perceived violations of the Wall Street Consumer Protection Act
Consumer Product Safety Improvement Act (CPSIA)	Applies to employees of consumer product manufacturers, importers, distributors, retailers, and private labelers
Energy Reorganization Act (ERA)	Applies to reporting of violations from employees of operators, contractors, and subcontractors of Nuclear Regulatory Commission (NRC) licensed nuclear power plants and employees of contractors working with the Department of Energy under a contract pursuant to the Atomic Energy Act
Federal Water Pollution Control Act (FWPCA)	Applies to an employee who reports violations of the discharge of pollutants into water
International Safe Container Act (ISCA)	Applies to violations reported by employees involved in international shipping who report to the Coast Guard the unsafe intermodal cargo container or another violation
Occupational Safety & Health Act (OSHA)	Applies to employees who file a safety and health complaint with Occupational Safety and Health Administration (OSHA)
Toxic Substances Control Act (TSCA)	Applies to employees who report violations relating to industrial chemicals produced or imported into the United States
Sarbanes-Oxley Act (SOX)	Applies to employees of publicly traded companies and employees of nationally recognized statistical rating organizations who report alleged mail, wire, bank, or securities fraud; violations of the Securities and Exchange Commission (SEC) rules and regulations; or violation of federal laws related to fraud against shareholders
Wendell H. Ford Aviation Investment and Reform Act for the 21st Century (AIR21)	Applies to employees of air carriers and contractors and subcontractors of air carriers who report violations of related to aviation safety

6.3 Elements and attributes of culture for peak performance

Culture research and theories have evolved and matured over the years as a result of extensive research and theories presented by many authors and practitioners.[1] Through this evolution, some key elements continue to surface as attributes of significant importance. There are six elements that are viewed as critical attributes of a healthy culture. These elements include openness and honesty, competence, reliability, identification, concern, and relationship. Each of these elements embodies the definition of culture and forms the basis for trust, values, beliefs, behaviors, and the practices seen in organizations. These elements are defined in Table 6.2.

Together these elements form the glue that can bind teams together and form a cohesive group of people that respect and support each other and have the desire to work together for the good of the team while sharing knowledge and ideas needed to propel a company toward achieving at its Peak Performance.

6.3.1 Trusting cultures

There have been many research studies conducted throughout the 1990s and 2000s that have linked high levels of trust with greater performance in organizations[1]. This research, along with many practitioners' experiences, has identified a trusting workplace as being instrumental to the success of organizations. Trust is viewed as a social expectation that deals

Table 6.2 Cultural Elements

Element	Description
Openness and Honesty	Communicate completely and fully information known
Competence	Leaders demonstrate their competence and their ability to hire and retain competent workers
Reliability	Leaders demonstrate that they are reliable in their actions (dealing with employees fairly, consistent and stable decision-making)
Concern	Leadership demonstrating concern for employees through their actions, policies set, and the channels used to engage workers in the business
Identification	Employees feeling like they can identify with the leadership team in areas such as having the same goals and organizational values
Relationships	Leadership and employees have established and are able to maintain an open, honest, and professional relationship

explicitly with the perception of people when it comes to integrity, honesty, caring, and competence of individuals or the system in place. Trust is highly dependent upon human relationships therefore; organizational situations and interactions can encourage or discourage an individual's ability to trust. In order to build and sustain trust, leaders must be aware of situations or actions that can lead to distrust so that these actions can be avoided. Some actions that can lead to distrust include:

- Management display of abrasive, ambiguous, and erratic behavior
- Employees having negative perceptions about the leadership team and the organization's culture
- The human resource policies and systems do not support an environment of fairness and respect

On the other hand, it is also necessary to know the actions that can lead to trust so that they can be practiced. These actions include:

- Having clear organizational goals, mission, values, and objectives, and demonstrating these tenets in management decision-making
- Being honest and forthcoming in all communications
- Demonstrating concern for the wellbeing of others
- Understanding the impact of individual actions on others
- Removing barriers that demoralize employees

Developing a culture where trust flourishes across the organization drives strategic leaders that are willing to chart the way and serve as examples of what they are expecting to achieve, and what they are expecting from every member of the organization. Trust has been linked to several positive organizations' characteristics such as increased productivity, retention of employees, creative thinking, investment in organizational success, and engagement of organizational members. Trust is also credited with increasing positive workplace interactions, such as positive attitudes and behaviors, higher levels of cooperation, better team interactions, enhanced workplace relationship, increased collaboration, and increased employee morale. By knowing the value of trusting cultures, Managers are encouraged to facilitate trust through all organizational processes and actions.

6.3.2 Learning culture

Training can be administered through various means such as in the classroom, on the job, shadowing, or through coaching. Many assume that just because training has been administered, learning has taken place. This assumption has been proven to be in error time and time again. In order for learning to take place the method of information delivery and the culture of the organization must be aligned. A learning culture typically has the values, policies, practices, and procedures in place that encourage organizational members to continue to seek and gain knowledge as well as learn from mistakes. These organizations encourage continuous improvement, strategic thinking, and innovation. Key benefits of learning cultures include:

- Increased productivity
- Enhancing the change management process
- Facilitating employee engagement
- Developing and maintaining trusting relationships
- Reducing or eliminating fear of failure
- Improved willingness for knowledge gaining and sharing
- Readiness to admit mistakes

It has been recognized that providing training does not mean that learning will take place *and* be sustained. Therefore, training should be developed and implemented with this knowledge in mind to be effective. The focus has now turned to facilitating cultures that are capable of enhancing and sustaining learning. Key elements often seen in learning cultures are shown in Figure 6.1 and discussed in detail in Table 6.3. In this type of culture, these components are visible and embraced by members of the organization, beginning at the Senior Management level. As such, the organization is able to perform at its highest level and is able to deliver superior employee development, products, and support to customers.

6.3.3 Safety culture

Every company strives for excellent safety performance for a variety of reasons. If workers are conducting work safely, incident rates are low, which means few worker compensation claims, less overtime to cover when an injured employee is out of work, and improved employee morale. The culture of an organization can be seen as important to the operation of the company, but also as key to the company's safety performance. Similar to the increase seen in production when workers have a high trust for the leadership team, the same is true for safety performance. When a company has a robust safety culture, the workers feel free to raise issues to the leadership team, including safety issues. Workers want and need to know that leadership cares about them as people, not just as objects that complete tasks. Managers want to have workers that care about their job and the condition of other employees – employees that watch out for each other's safety.

To build trust from a safety perspective, Managers must listen to employee's safety concerns, ensure work instructions and procedures reflect requirements for proper personal protection equipment (PPE), ensure adequate and proper PPE is available, and finally reinforce the

Chapter six: Culture

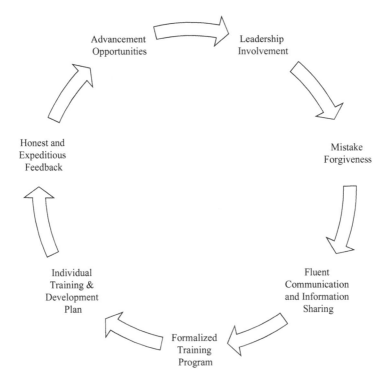

Figure 6.1 Key components of learning cultures.

expectation to properly wear and utilize PPE. Managers must observe work being conducted to ensure that safety is discussed as part of a pre-job briefs, the work area is safe for workers to conduct tasks (i.e., excavations are properly barricaded, scaffolding is properly erected, etc.), and finally to ensure workers are safely conducting tasks (i.e., not walking under cranes with suspended load, observing barricades, etc.). Companies that have a strong safety culture typically have very low incident rates and employees that want to work for them because they are confident that they will go home to their families in the same condition that they came to work.

6.4 Role of leadership in culture development and sustainment

Leaders play important roles in all activities within an organization, from strategy development, policy enactment, championing change, process improvement, and facilitating culture. These roles should not be understated or under-estimated, as it forms the basis of the culture of the organization as well as the path an organization takes to success and their ability to achieve their Peak Performance. Leadership is about having the

Table 6.3 Learning Cultures Components

Component	Description	Challenges
Leadership Involvement	Leadership sets the expectations and models the behavior that demonstrates their support and value for a learning culture	Leaders typically spend a significant amount of time with new and prospective clients strategizing, and with the public, as such may neglect being intimately involved in the internal business
Mistake Forgiveness	Mistakes that are made while following company procedures and policies should not be held against workers. These mistakes should be used as "teachable" moments so that organizational members can learn from them	Being able to balance or determine what actions are merely mistakes versus a disregard for following policies or procedures
Fluent Communication and Information Sharing	Free-flowing accurate communication that flows at all levels is necessary for creating trust and teamwork	Accurate and timely communication is not always easy when it needs to be viewed by different people or organizations for contentment and continuity
Individual Training and Development Plan	Training plans targeting gaps of specific individuals to prepare them to take on key positions	Requires leadership commitment to support resources needed to close gaps in time when financial resources may be short
Formalized Training Program	Documented training program that is evaluated for effectiveness at some frequency to ensure that the program is meeting the needs of the company and employees	Formalized training program is not easy to maintain. Training is generally one of the first areas that is cut when finances are of concern
Honest and Expeditious Feedback	Provide honest feedback to workers as soon as the information is known and avoid or reduce the rumor mill spreading information that is not complete or accurate	There are times when leadership may need to digest the information before a decision is made and the information can be communicated. The time taken may not be expeditious enough to avoid concern or prevent the rumor mill from developing

(*Continued*)

Chapter six: Culture 137

Table 6.3 (Continued) Learning Cultures Components

Component	Description	Challenges
Advancement Opportunities	Employees are interested in career advancement and adding more funds to their paychecks. They tend to work longer for companies that are able to fulfill these two requirements	Internal advancement is at times difficult for Managers, as they must develop and implement a strategy to replace employee transition, and train new employees, while maintaining production

ability to get results from people that support the value of the organization, mission, and business strategies.

Leaders are generally looked up to by workers and workers will typically emulate the behaviors and sanctions of leaders. As such, leaders have the ability to shape an organization through their prevailing behaviors. Therefore, leaders should be cognizant that they are impacting the culture of their organization with everything they say and do.

6.4.1 *Followership in workplace cultures*

In a progressive and optimized workplace culture, workers are willing to follow their leaders, organizational policies, procedures, and practices. People will only follow when certain organizational characteristics are presented and supported. These characteristics include an engaged and respected leadership team, policies that are viewed as being equitable and fairly implemented, the ability of workers to feel free to become and remain actively engaged. Each of these characteristics works together like gears in a system that are instrumental in keeping the process functioning to its optimal efficiency, as shown in Figure 6.2.

Any one characteristic of the Follower Paradigm can cause an imbalance in an organization's productivity and cohesiveness in implementing policies and procedures. In order for leaders to inspire people to become followers, they themselves must possess specific characteristics that will cause people to want to follow and emulate their actions. Often leaders are careless with their roles as leader and display actions that can hamper the progress and success of an organization. When this occurs, they are not fully aware of their actions and reactions during communication, verbal as well as nonverbal. The actions that a leader must take to inspire people to follow include:

- Walking the talk. Leaders must behave in the manner they expect other members in the organization to behave. The phrase "do as I say and not as I do" when observed in the leadership team can erode organizational systems that can include the development and sustainment of trust

- Demonstrating concern for the organization and its members. The best way for Managers to show that they value and are concerned about workers is to implement policies and procedures that serve the needs of the employees, as well as the organization. These policies and procedures must be implemented fairly and everyone must be held accountable for following them
- Support and facilitate a culture of inclusiveness. People respond well when they feel that they are a part of the cultural environment and that they are valued. This means equal treatment of organization members is a must
- Act with purpose and competency. The leader must demonstrate through actions that they have a strategy for the organization and that they are competent leaders

The characteristics listed in Figure 6.2, coupled with the appropriate actions of the leadership team, create an environment for followers to follow willingly to ensure that the goals of the organization are met.

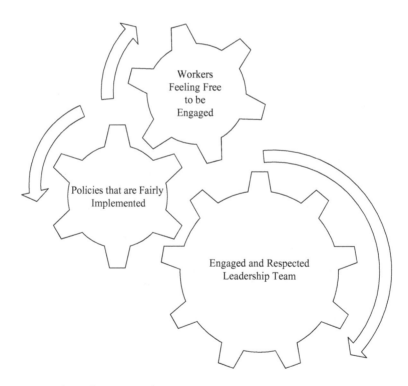

Figure 6.2 The Follower Paradigm.

6.5 Evaluating the health of organization culture

There are a few methods that are currently being used to measure or evaluate the health of the culture of an organization. These methods include evaluating the content of policy and procedures, administering a survey instrument, interviewing individual organization members, conducting focus-group discussions, and conducting work observations. Collectively these methods can shed light on the health of a culture and areas that should be targeted for improvement.

Culture has been extensively studied for many years, so there are many survey instruments available to measure it. These instruments differ in part due to variations in definitions and views of organizational culture by theorists and practitioners. Among the first recognizable survey instruments used to measure culture characteristics is the Profile of Organizational Characteristics (POC) developed by Likert.[2] The POC survey instrument consists of 16 questions that are used to measure culture. The POC survey instrument was developed to measure management styles and organizational characteristics that include leadership, motivation, communication, decisions, goals, and control mechanisms. The POC instrument was validated initially in 1964 during the General Motors study.[2] Other culture survey instruments were developed in later years.

In recent years, culture is being assessed using more comprehensive means, which include surveys (examples shown in Tables 6.4 through 6.6), employee interviews (example interview questions listed in Table 6.7), work observations, and documentation outlining the expectations for the way an organization operates. Another key aspect of evaluating culture is having knowledge of whether trust among organizational members is at an acceptable level. As a part of evaluating culture, it is advisable that organizational trust is evaluated at the same time. Organizational trust can be evaluated using methods similar to those one would use to evaluate culture, with the exception of employing a survey that can specifically evaluate trust (Table 6.5) and focus interview questions specific to trust.

6.5.1 Policy and procedure reviews

The first step in evaluating culture should begin with a review of key policies and procedures that set the stage and outline the expectation for the way the organization members should behave, operate, and their expected practices. In organizations where the policies and procedures are not supportive of the vision and the expectation for behaviors are not set, workers are confused about what is expected and will do what they believe to be appropriate.

Table 6.4 Organizational Culture Questionnaire[1] (Alston, 2016)

Organizational Culture Survey

For each question given below, circle the number that best describes your opinion to the questions listed

1	2	3	4	5	6	7	8
Always	Mostly	Frequently	Usually	Sometimes	Infrequently	Seldom	Never
			Question				
1	My organization has a clear vision						
2	The values of my organization are shared by its members						
3	I believe that my management values my opinion						
4	Communication in my organization is fluent and flows in all directions						
5	The Managers in my organization recognize and celebrate the success of its members						
6	The mission and values of my organization are posted for employees to view						
7	In my organization management celebrates the successes of employees at every level						
8	The management team is trusted and respected by employees at every level						
9	Management is responsive to suggestions from employees						
10	Conflicts are handled openly and fairly						
11	Employees are motivated to perform their jobs						
12	Employees understand their job duties and their role within the organization						
13	Downward communication is accurate						
14	The organization goals and objectives are clear to employees throughout the organization						
15	Roles and responsibilities within the organization are clear and understood						

(Continued)

Table 6.4 (Continued) Organizational Culture Questionnaire[1] (Alston, 2016)

Organizational Culture Survey

For each question given below, circle the number that best describes your opinion to the questions listed

1	2	3	4	5	6	7	8
Always	Mostly	Frequently	Usually	Sometimes	Infrequently	Seldom	Never
			Question				
16	My input is valued by my peers						
17	Employees have the right training and skills to perform their jobs						
18	Knowledge- and information-sharing is a common practice for members of my organization						
19	Disagreements are addressed properly when they occur						
20	Morale is high across my organization						
21	Employees enjoy coming to work						
22	I feel that I am valued as a part of my team						
23	Employees speak highly of my organization						
24	Everyone takes responsibilities for their actions						
25	My Supervisor is a positive role model						

Survey Average: _____

Table 6.5 Example Organization Trust Questionnaire

Question#	Question	1	2	3	4	5
1	Senior Management in my company communicates information completely and frequently					
2	My Supervisor communicates information frequently and accurately					
3	The Managers in my organization keep promises					
4	I feel safe while working in my workspace					
5	The people in my organization always follow procedures					
6	Clear concise communication is seen throughout the organization					
7	People in my organization treat each other with care and respect					
8	Communications flows in all directions in my company to ensure that all workers are kept informed					
9	Disagreements and issues are addressed in a timely manner					
10	My Managers support a work–life balance for workers					
11	Managers in my organization openly admit mistakes when they occur					
12	Leaders in my organization are decisive decision-makers					
13	Management always communicates openly and honest					
14	Workers are actively listened to by management					
15	Employee benefits in my company are comparable to similar companies					
16	Information is freely and willingly shared by organization members					
17	My management team leads with confidence					
18	Policies are in place to ensure employees are treated fairly across the organization					
19	Managers in my organization demonstrate good leadership and management skills when conducting business and making decisions					
20	The Managers in my organization are viewed as being competent leaders					

Instructions: Place an X in the space that has the number that best describes your opinion of the questions being asked below. Only one number should be selected for each question and please respond to all questions.

1 – Always; 2 – Sometimes; 3 – Usually; 4 – Seldom; 5 – Never

Table 6.6 Example Employee Engagement Questionnaire

Question#	Question	1	2	3	4	5
1	My opinion is often sought by management					
2	My coworkers often ask for my input					
3	I willingly participate on teams to develop and implement new technology, programs, or processes					
4	I share responsibility for safety of myself and my coworkers					
5	I feel valued for my contribution to the success of the organization					
6	My immediate Supervisor provides guidance on how to perform work					
7	I have the opportunity often to participate in providing input on decisions affecting me and my team					
8	I willingly provide feedback to management on issues					
9	Employees in my organization are motivated to accomplish work					
10	My management team inspires me					
11	I enjoy going to work					
12	Employees in my organization are accountable for their actions					
13	I am proud to tell others where I work					
14	I am provided the tools I need to effectively perform my task					
15	I know and understand how the work that I do supports the goals of the company					
16	I am committed to doing a good job					
17	I would happily recommend my organization as a good place to work					
18	I know what my Supervisor expects of me					
19	Employees in my organizations often proactively offer solutions to problems to management when issues are discovered					
20	I am satisfied with my job					
21	Management encourages worker involvement in organizational activities					

Instructions: Place an X in the space that has the number that best describes your opinion of the questions being asked below. Only one number should be selected for each question and please respond to all questions.

1 – Always; 2 – Sometimes; 3 – Usually; 4 – Seldom; 5 – Never

Table 6.7 Example of Interview Questions

1. What do you enjoy most about working in your organization?
2. What words would you use to describe the culture of your organization?
3. What are the most common complaints workers have concerning your organization and leadership team?
4. How do communications flow in your organization?
5. Are you provided the equipment and training needed to perform your job safely and successfully?
6. How comfortable are the people in your organization in communicating with management (Supervisor, Middle Management, Senior Manager)?
7. How comfortable are you in communicating with management (Supervisor, Middle Management, Senior Manager)?
8. How comfortable are you in raising concerns to management?
9. Are the skills and abilities of the people in your organization valued by management and used in your current job?
10. How important is trust to the employees in your organization?
11. Are employees encouraged and willing to get involved in solving issues?
12. What words would you use to describe the management team?
13. How would you describe your Supervisor's leadership abilities?
14. How does the leadership team demonstrate trustworthiness?
15. How often do you communicate your ideas and suggestions on improving business processes?
16. How do the people in your organization handle changes?
17. How engaged are the employees in your organization when it comes to participating in business outcomes?

Documentation reviews can also be the final step in the evaluation process. One may be wondering why there is a need to review company policies and processes once the assessment has been completed. A review may be necessary, to determine if the results of the assessment need to be incorporated into company policies or procedures. The culture assessment process is shown in Figure 6.3.

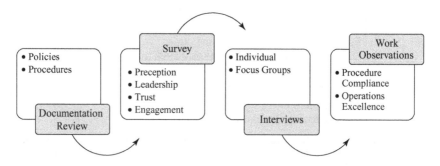

Figure 6.3 Culture assessment process. Note: Document review can be completed as step one or as the final step in the process.

Chapter six: Culture 145

6.5.2   *Focus group and individual interviews*

A focus group can consist of a group of people of around 6 to 12 participants. The objective is to keep the group small enough so that free flow of communication and information is facilitated. Focus group discussions are a valuable tool used to learn more about the opinions of organization members on designated topics. Some guidelines to consider when conducting focus group discussions are:

- Keep the list of questions short, in the range of 10–15 questions, and use the same question-set for all the interviews
- Ensure questions are short and clear
- Ensure the use of terms that are familiar to the participants
- Use open-ended conversational-type questions
- Avoid using leading questions
- Ensure that the questions are not threatening
- Each question should focus on a single element or activity
- Select participants that have knowledge of the topic
- Select the appropriate meeting location that would ensure privacy for the group

Individual interviews can be used when gleaning information from one individual. This method is the best method to use to get feedback from Senior Managers. The same questions used during the focus group discussions should be discussed during the interview process, along with additional questions that may be appropriate only for Senior Managers. Individual interviews should be kept to the prescheduled timeframe. The use of focus groups and individual interviews are a good combination to collect additional data on the cultural health of an organization. Together, the information gleaned can provide valuable insight into the views and opinions of the leadership team and the workforce.

6.5.3   *Culture sustainability strategy*

Many companies are beginning to include in their overall business strategy provisions to ensure that the culture of the organization is evaluated and that impactful changes are made as needed.

A comprehensive culture improvement strategy contains provisions to:

- Determine the status or health of the organization's culture through the use of survey instruments, focus groups, or individual interviews
- Determine if the company procedures and policies are aligned with the culture

- Employ metrics to determine progress, identify problem areas, and evaluate the success of corrective actions after they are implemented

A strategy for culture measurement and enhancement is most effective when included in the overall organizational strategy.

References
1. Alston, F. (2014). *Culture and Trust in Technology-Driven Organizations*. Boca Raton, FL, CRC Press.
2. Likert, R. (1967). *The Human Organization: Its Management and Value*. New York, McGraw-Hill Book Company.
3. Schein, E. H. (1992). *Organization Culture and Leadership*. Hoboken, NJ, John Wiley & Sons.

chapter seven

The peak performance model

7.1 Introduction

All companies have ups and downs in every facet of business, from production to profit-margin challenges. The causes of these challenges may vary, but the ability of a company or institution to identify and adapt to changes will determine if they are able to achieve improved operational performance. The difference between a company that performs marginally and a company or corporation that achieves and sustains Peak Performance is the ability of a company or institution to analyze the issues and develop a comprehensive improvement plan. Both Management and employees must understand the issues, be involved in the analysis, and, most importantly, be involved with the development of solutions. Everyone must own the issues and work in concert to resolve them. The team must be able to recognize the importance of resolving issues quickly, but with care, to ensure that the corrective actions are effective and sustainable. If the team is functioning in an environment that has a strong culture and where members trust each other, the tough times will not seem as rough and/or perhaps as insurmountable.

Productivity is directly associated with facility operability, the caliber of the leadership team, the culture of the organization, and employee engagement. For example, to ensure production rates are sustainable, equipment must be maintained, preventive maintenance completed as prescribed by the manufacturer, and corrective maintenance completed when required. When a Leadership Team does not maintain equipment or facilities it can send a message to the employees that it is permissible to use workarounds or take short cuts. To be a successful company, equipment maintenance is a priority, and equipment operability should be tracked and trended to support maintenance needs and the recommended replacement schedule. A successful maintenance program should be linked to required outage timing, and the type and number of spare parts needed to maintain the equipment. A well-planned facility outage reduces the time the equipment is offline because the work packages are prepared in advance, spare parts have been ordered and arrive at the facility in time for maintenance to occur, procedures written, and employees trained if necessary. With all of the elements ready for an outage, work can be completed in an orderly manner, using operations excellence and Human Performance tools for a disciplined approach.

Another key to maintaining productivity is employee engagement and leadership involvement. When employees are engaged, issues are brought to the attention of the Management Team and feedback is provided. For example, if a motor is making an unusual sound, you want your employees to tell someone from the Management Team. Then you would want the Management Team to respond, first with a thank-you for bringing this issue to their attention and then, second, you want the manager to follow up on the concern. Consider what it might be like if all your employees acted as if they owned the equipment, and they were constantly looking for ways to improve the product and/or reduce cost.

So, let's look at how one can evaluate a company's progress toward achieving peak performance. As discussed in Chapters 2 through 5, there are three continuous improvement processes which many industries use individually to achieve improved performance. These processes are:

- Lean thinking
- Human Performance Improvement
- Operations Excellence

These processes and programs are well established and have been implemented in the commercial sector and within government agencies for many years. So how can a company, or institution, that implements one or more of these continuous improvement processes achieve even greater efficiency and improved performance? What tools can they use to achieve greater efficiencies with limited funding and resources? What is Peak Performance for an organization, institution, or company?

Peak performance occurs when a company or institution exhibits:

- A robust organizational culture where employees feel valued and appreciated
- Processes are streamlined and employees want to follow policies and procedures
- Waste is reduced and rework almost eliminated
- Human error is anticipated and managed to reduce process upsets, all while maintaining a safe working environment

This is a work environment where employees want to come to work and focus on improving not only their own performance but also want to contribute and promote the organization's or company's success.

So, what is the Peak Performance Model and how can it improve operational performance? It is a continuous improvement model, which integrates three common continuous improvement programs: Lean, Human Performance, and Operations Excellence to optimize application of shared attributes. In addition, the Peak Performance Model recognizes

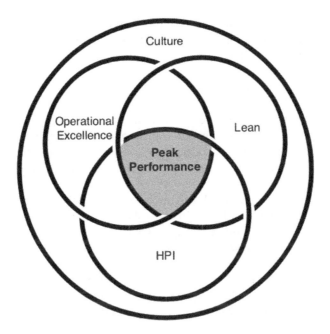

Figure 7.1 Peak Performance Model.

the influence culture has on each of the programs and how the health of the culture can influence the success of these continuous improvement programs, the organization, and the company. Organizational culture is the fourth functional element of the model. Figure 7.1 is a diagram of the Peak Performance Model, which depicts the four functional elements and their interrelationships.

7.2 Lean

Lean is a way of thinking that leads to instituting processes and practices that focus on creating value for the customer through the elimination of waste. System and process waste tends to create additional cost for a company that is generally passed on to their customers. Many companies leverage Lean as a way of thinking to help an organization increase in efficiency and provide better value to the customer.

As discussed in Chapter 2, implementing Lean requires an understanding of how company and operational processes work, a Lean-focused leadership team, and engaged employees that are able to think outside of the box. All across the United States, companies are using the Lean Process, Six Sigma, or a combination of Lean Six Sigma to target and identify unnecessary waste, or tasks being performed that add no value to the customer and decrease business efficiency (along with profits). In a company operating at Peak Performance, the major processes have been evaluated to eliminate waste, reduce wait-times, and obtain the necessary supplies just in time, not stocked in some warehouse adding no value.

7.3 Human performance improvement

All companies need employees to sustain business and operations. We also know from various studies that humans are fallible. Errors committed by humans can impact productivity, worker safety and health, product quality, long-term sustainability, and overall financial burden to the company.

As discussed in Chapter 4, the purpose of Human Performance tools is to help workers gain and sustain control of a process or work condition, such that what is envisioned to happen actually happens, and that is all that happens. Another way of saying this is, "Do the job right the first time." The tools presented in Chapter 4 can help workers reduce errors and are actively utilized in a Peak-Performing organization. When operating at Peak Performance, learning and mastering the Human Performance tools will support error reduction. The Human Performance tools give workers more time to think about the actions they are about to take, what is about to happen, and what they can do if things do not go as predicted.

In a company operating at Peak Performance, mistakes are used as learning opportunities for the workforce, tasks are previewed before they are conducted, to look for potential error precursors, critical steps are identified, and tools are implemented to reduce the possibility of an error at each critical identified step.

7.4 Operations excellence

Operations Excellence instills rigor in performing operations to ensure product quality and reliability. While Lean focuses on the needs of the external customer, Operations Excellence focuses on internal customers and processes that support efficient and robust operational performance.

As discussed in Chapter 5, Operations Excellence is visible when communication flows at all levels and in all directions in the organization, workers clearly understand the tasks they are asked to perform, and the leadership is actively engaged in all facets of the business. Workers feel free to communicate, plan, participate, and provide feedback on work activities. The organization utilizes a variety of tools to constantly measure and improve performance. In a peak performing company, workers know the expectations for tasks being performed, feel free to communicate issues openly, and leadership is engaged with the workforce to promote and maintain a culture of trust. Peak performing companies recognize the importance of focusing on both internal and external customer needs to ensure the best-quality product is developed, produced, and delivered.

7.5 Organizational culture

The culture of an organization serves as the foundation upon which employees interact, form relationships, and support business processes and practices. In organizations that use the Peak Performance Model, cultural attributes such as common practices, values, beliefs, behaviors, languages, and symbols are visible and used to unite workers to serve a common goal.

As discussed in Chapter 6, these attributes serve as the foundation for the cultural health of the work environment to ensure that Peak Performance can be obtained and sustained. In addition, these cultural attributes are used to build relationships and instill trust in organizations.

In a company functioning in the peak performance realm, workers and the leadership team have shared visions and values, workers feel free to openly share issues and express their point of views, trust the message delivered by the leadership team, and believe Management will follow through with commitments. In addition, Leaders are concerned about the wellbeing of the employees, and workers are engaged and support the goals of the company because they feel a sense of ownership and partnership with the company.

Also, companies and institutions implement their processes and programs in the most efficient manner, while promoting and maintaining a healthy work environment by individually, and collectively, evaluating the functional elements of the Peak Performance Model.

7.6 Integration of peak performance elements

When used effectively, the Peak Performance Model offers the opportunity for any company or institution to improve and sustain improved performance across the board. Each of the three continuous improvement processes represents a functional element of the model. The fourth functional element of the model is organizational culture because of the significant impact it has on the success of an organization or company through the manner in which employees function and the way work is accomplished. Each of these functional elements is integrated to provide optimum results, and all four of the functional elements share several common attributes. The shared attributes identified in the Peak Performance Model in many instances are also reflective of a companies' core value. These attributes are visible across the organization and have been assimilated into the model and peak performance model assessment tools.

Table 7.1 Individual and Shared Attributes of the Peak Performance Model

Key Attribute	Lean	HPI	Organizational Culture	Operations Excellence
Communication	X	X	X	X
Customer Focus	X	X		X
Process Improvement	X	X		X
Employee Engagement	X	X	X	X
Leadership Involvement	X	X	X	X
Trust	X	X	X	X
Shared Vision	X	X	X	X
Quality	X	X		X
Eliminate Waste	X			
Operational Rigor	X	X		X
Error Reduction	X	X		X
Problem Solving	X	X	X	X
Work Performance	X	X	X	X

The breadth and the depth of these shared attributes are evaluated, yielding a comprehensive review across the organization; all attributes are evaluated both from within, and across, the company. For example, leadership involvement is vital in all four elements of the Peak Performance Model. Assessing leadership involvement as part of each functional element provides a cross-functional view of leadership both within and across the organization. Leadership involvement is just one example of a shared attribute. Table 7.1 lists commonly identified desired attributes; attributes that are shared amongst all four of the Peak Performance Model functional elements are highlighted. The shared attributes have been incorporated into the peak performance model assessment tools to evaluate how well these attributes are being embraced and implemented by management and the workforce.

7.7 Application of the peak performance model

Application of the Peak Performance Model starts by evaluating each of the functional elements using the example assessment tools provided in each chapter. If your company has not implemented or used one of the three continuous improvement processes outlined in the Peak Performance Model, you may still assess your organization and use the assessment tools provided in this book to help gauge where your company is on streamlining processes, eliminating waste, reducing errors, or conducting disciplined operations. Cognitive decision-making is the primary means by which the Peak Performance Model is implemented. Remember to be self-critical, honest with the review and your responses; the only way to improve is to recognize the shortfalls. Take the time to observe primary mission-critical operations and processes, all shifts, and note if standard processes are utilized. Example questions that may be asked when evaluating the four functional elements are:

- Do each of the shifts perform the operation/task the same way using procedures?
- Looking around the facility/plant, do you see waste? Remember, waste can be product/materials or employees waiting for a piece of equipment, or walking to get parts, etc.
- Do employees know the goals of the facility/plant and are these goals visible to the organization?
- How do employees communicate with each other, is it clear and concise?
- What is the culture in the facility/plant, do the workers freely bring up issues, make recommendations, do they interact with the Management Team?
- How many errors are made on a daily basis, do the employees report when a procedure cannot be followed, do Managers discusses errors, critical steps, or error precursors?

These important questions, among others, need to be answered to really determine how the organization is performing. When conducting the evaluation, observe as many areas of the facility/plant as possible, including the administrative areas. The following are suggested items to consider when performing an assessment:

- Take detailed notes when observing operational areas.
- Talk with multiple workers at various sections, and various shifts of the facility/plant, to ensure an adequate sampling to inform the basis for your opinion.
- Take a good look around the facility/plant, note what you see. What type of conversations did you hear, what messages and posters are hanging on the walls, number of interactions between managers and workers, did you see workers using procedures, having discussions, etc.?
- Observe pre-job briefings, plan-of-the-day meetings, safety meetings, or whatever type of meetings your organization conducts, look for the tone of the Management Team, the questions being asked by the workers and the responses from the Management Team. Are the workers engaged or just sitting back, how do the workers respond?

- Observe workers and determine if they are seeking ways to make the operation better. Are workers willing to participate, or just to collect their paychecks and go home?
- Observe the Management Team and determine if they are interacting with the workers, involving the workers in decision-making, discussing how to make improvements, asking for feedback on how a task went. Were there any issues, errors in the documents, etc.?
- Talk to the employees and find out if they trust and believe the Management Team.

These types of observations will support an overall assessment and help form the opinion of how your facility/plant employees really feel about the Management Team. Observations will assist in understanding how employees feel about being part of the team, and support understanding how the Management Team interacts with the workers and involves them in decisions, listens to concerns and issues, and follows up on their concerns. Data generated is documented and organized by four shared attributes (the assessor can either select these or other shared attributes):

- Leadership
- Employee engagement
- Organizational factors
- Work planning and execution

Carefully review each assessment and determine what the overriding themes, or clear messages from each assessment, were. What did you believe you would find, what surprises did you uncover, was work completed as you thought, was the operation running as smoothly as you envisioned? The answers to these questions will probably surprise you. Like most Managers, companies want to believe their team is the best at what they do, and that very little improvements are really necessary. However, like most companies, opportunities for improvement are abundant; all one has to do is to take a self-critical look, and be honest with what was really going on in the facility/plant. Now that you have a clear picture of how your company really operates, it's time to develop and implement your plan to achieve Peak Performance.

7.7.1 Quantitative and qualitative evaluation of peak performance model functional elements

Each of the functional elements (Lean, Human Performance Improvement, Operations Excellence, Organizational Culture) are qualitatively evaluated using the example assessment tools provided in the respective chapters.

Data generated is documented and organized by four shared attributes: leadership, employee engagement, organizational factors, and work planning and execution. The Peak Performance Model can also be applied and data analyzed using other shared attributes and by use of quantitative methods. The quantitative analysis offers the following advantages over the qualitative analysis:

- Focuses on issues within each of the functional elements, thereby resulting in an efficient and effective improvement plan
- Reduces the introduction of subjectivity and bias of the assessors
- Direct, quantitative comparison of performance among Corporate entities or product lines and/or manufacturing facilities
- Tailoring of analysis to desired emphasis of end-state improvements
- Enables lessons learned to be horizontally applied across the corporation, company, and/or organizations

Through the use of a quantitative analysis method, companies and organizations can perform a direct comparison of improvement actions to support prioritization of funding and future planning.

Whether a company or organization evaluates data using the qualitative or quantitative method, the analysis should consider company and Management goals and objectives, behaviors exhibited by Management and the workforce, and whether behaviors, beliefs, and leadership are driving performance in a positive and healthy manner. It is also recommended that Management consider utilization of a team of people to perform the analysis so that different perspectives and viewpoints associated with data analysis can be represented. It is best performed by a multidiscipline group with skill sets that range from being an individual contributor, craft personnel, and Management.

7.7.2 Evaluation of peak performance model functional elements

Once the assessment has been completed, generate a list of the issues that were observed for each functional element, noting whether the issue was systemic or localized, and also note whether the issue was expected, based on the procedures and policies that are in place. Organize the list by selected shared attribute: leadership, employee engagement, organizational factors, and work planning and execution. This will support the development of the overall improvement plan.

In each of the four attributes prioritize the list from most important to least important for improving your business. Data within each of the four shared attributes is then analyzed, using cognitive decision-making, and an overall analysis of the health of the functional element can be determined. Results of the assessments can be used to identify, prioritize, and target funding and corrective actions to improve implementation of each Peak Performance Model Functional Element.

7.7.3 Collective evaluation of peak performance model

Once each of the functional elements have been individually evaluated, the functional elements are then evaluated by comparing and analyzing not only results of the individual analysis of each functional element but, collectively, comparing the effectiveness of the three continuous improvement program and organizational culture to determine whether there are common themes that represent strengths or weaknesses across all functional elements.

By comparing assessment results from all four functional elements, Management can then further target and focus the application of resources and funding on improvement actions that can achieve greater impact and efficiency of labor and dollars. In addition, those attributes that are common among all four functional elements are further evaluated to determine if the weakness is systemic and a larger realignment of the continuous improvement programs are needed. Under each element, there should be a list of items for targeted improvement based on common attributes and how they would apply to the functional element as part of the model. An operational improvement plan can be developed.

When comparing collective qualitative analysis results to quantitative results, several benefits emerge for quantitative application including:

- Provides a company an independent appraisal of their performance in relation to continuous improvement processes and organizational culture
- Reflective of the overall health of the company or organization and whether the behaviors associates with the shared attributes are being exhibited to the degree desired by the company
- Results in a numerical comparison between continuous improvement program and culture
- Improvement action items are based on overall prioritization of continuous improvement program needs

7.8 Development of a peak performance improvement plan

The plan to achieve Peak Performance should address the challenges with comprehensive solutions, going after the tough issues, because that's the only way your business will achieve and maintain the performance you are striving to obtain. The plan must be achievable, not so lofty that no one buys into the solution. Most companies can only realistically work on 8 to 10 complex improvement actions at a time. So, going back to the prioritized list, select the number of issues that you believe are achievable within a 6- to 12-month period of time.

Solutions that will address the issues should be credible, comprehensive, and implementable. Conduct root-cause analysis if you are struggling with what caused the issues, remembering to involve the workforce, since the workers are the ones that know how work is carried out throughout the facility. The improvement plan should address:

- Targeted area for improvement
- Actions needed to achieve improved performance
- Responsible party for ensuring completion of the improvement action
- Desired timeframe for completing the action
- Performance measure to monitor completion of action and improvement activity

The final step of the plan is to develop and implement a communication strategy to inform the workers why the plan is necessary, what has been done, and why. The plan should be published and made available to all employees, to begin soliciting buy-in for the actions. Employees should have an opportunity to provide feedback on the plan, as it might generate additional actions or some actions may need to be clarified, and it provides an opportunity for employee engagement, which is a good thing. At this point, metrics should be developed for each improvement action, visibly tracking the status of each action for the entire facility/plant to see, with the results posted. This sends a message to the workers that continuous improvement is a core value for the organization and that the Management Team holds itself accountable.

Once all actions have been completed or implemented, reassess progress, and add additional improvement actions based on the next set of issues. After completing all of the actions associated with the issues identified, and allowing time for the improvement actions to be engrained in the company work processes, reassess the facility/plant. To gauge improvement, Management and employees should ask the following questions:

- Did the actions completed improve the culture?
- Did the actions streamline the process, reduce waste, and eliminate rework?
- Did the actions reduce human errors?
- Did the actions significantly reduce events/incidents (in comparison to operational performance over previous years)?

If your business did not reach the desired performance, reassess the operation and develop another improvement plan. Keep completing the cycle until you achieve your Peak Performance.

7.9 Journey to achieving and sustaining peak performance

In order to sustain a thriving business, it must be constantly assessed and monitored. Performance metrics are one method for continuously monitoring performance; however, other methods can be used as well. The assessment process reveals actions that need to be taken to improve those areas that have slipped and those areas that are still not meeting expectations. All areas of the business should be assessed and monitored. Remember to use the "Lean Lens" when conducting facility walk-throughs, looking for opportunities to reduce the waste in the process and trend performance over time.

Ensure workers are following the procedures and not utilizing work-arounds. The Management Team is encouraging the use of Human Performance tools and setting expectations for operating in a disciplined manner. Probably the most challenging and important attribute is to ensure a strong safety culture is cultivated and maintained. The Management Team must develop and sustain trust with the workers and encourage open and honest communications and provide feedback when necessary.

It is important not only to assess your performance but to monitor your performance. Often companies utilize a variety of metrics to monitor and identify positive and negative operational and business performance. Table 7.2 provides a list of examples of key metrics that can be used to monitor and trend Peak Performance.

Once your company has internalized the three continuous improvement processes, understanding the value that each brings to the overall success of the company, along with the importance of a strong organizational culture, your Leadership team is positioned to enjoy the benefits of moving forward and achieving Peak Performance. It takes time to achieve Peak Performance, but through application of the Peak Performance Model, a company can go from failing to leading in the markets in which they compete and can achieve excellent operational and business performance.

Table 7.2 Example Key Metrics

Metric	Expected Outcome	Examples of How to Measure
Employee Engagement	Trending involvement in all facets of the business to include problem solving	• Survey • Number of Lean/safety suggestions • Employee satisfaction survey • Employee participation in employee-owned committees
Production Goals	Trending performance overall time – expect goals to be met	• Weekly attainment versus goal • Maintenance backlogs • Operational time
Safe Performance of Work	Trending performance over time – expect to see improvement	• OSHA reporting – lost work-day cases, etc. • Number of first-aids reported • Near-miss events
Leadership Involvement	Trending time in facility per manager; keeping abreast of processes and work practices	• Number of hours per month in facility • Survey • Group interviews • Attendance at employee-owned committees • Number of All-Hands Meetings
Lean Process Improvement Projects	Evaluate key processes to eliminate waste. Operations will become more efficient and effective	• Number of Lean events • Employee participation in Lean events • Actions completed • Cost reduction
Number of Operational Incidents	Trending incidents over time to determine frequency of occurrence	• Number of incidents over time • Number of error precursors identified over time
Customer Satisfaction	Trend over time to ensure customer sustainment	• Number of complaints • Voice of Customer interviews
Errors Realized/Reported	Trending errors over time to determine frequent and cause of errors	• Number of errors per work hours • Number of tasks completed without errors per month
Procedure Compliance	Trending number of procedure unable to be completed as written	• Number of procedure non-compliance per month

chapter eight

Practitioner's guide case study
Spud's Chemical Company, LLC

Spud's Chemical Company, LLC, located in Pacer, North Carolina, manufactures lacquers for furniture finishes. The company was family-owned for 45 years until the D.O. Chemical Company purchased it two years ago, consolidating the companies and creating a new subsidiary of the Corporation. Spud's Chemical Company has historically employed around 150 employees that annually produce over 1,500,000 gallons of lacquer. The workforce is comprised of 75% very experienced workers (greater than 20 years at the company) and 25% with less than 2 years' experience; the age of the average worker is 52 years. Since the acquisition of Spud's Chemical Company by the D.O. Chemical Company, production capability of the Spud's Chemical Company facility is being expanded, and operational equipment upgraded to support an increase in the facility production capability to a 24-hour-a-day, seven-days-a-week production schedule. Ten new production units have been designed and are currently in the process of being installed and tested for startup. Figure 8.1 shows the new facility layout for each of the ten units that will produce a variety of lacquers.

Until a month ago, Spud's Chemical Company only operated on day shift. Recently, as design, construction, installation, and readiness of the ten new units were nearing completion, Spud's Chemical Company announced that they would start placing workers on four shifts to support a 24-hours-a-day operation to meet increased production goals.

The new shift assignment created difficulty for some workers. The workers were assigned into four rotating shifts, and work assignments were made such that many workers were assigned to work with crew members with whom they had never worked before.

A shift consisted of one Shift Manager, five Team Leads, 20 Operators, two Maintenance Mechanics, and two Electricians. The Management Team did not request input on how the workers would be placed on shift, or who would be appointed as Managers and Team Leads. Some of the newly appointed Managers and Team Leads had never functioned in a leadership role and were not given any training prior to their new appointments. They were selected solely based on their service time with Spud's Chemical Company. The employees were not happy with the new

Figure 8.1 Facility layout for each of the new lacquer units. (Courtesy of PRESENTERMEDIA.)

requirement to work shifts and the Management Team did not really try to address any of the safety concerns voiced by the employees. For example, over the past two years, several workers expressed their frustration with accessing and operating several manual valves to enable proper alignment on existing plant equipment. Just last month one of the operators was standing on a ladder and tried to reach a valve on an existing storage tank when the operator missed the valve with his hand and ended up hitting his elbow on the side of the tank. Luckily, no one was seriously hurt, but the operators had not been happy with how the tanks and existing systems were installed and had repeatedly expressed their concerns.

Prior to being employed at Spud's Chemical Company, several of the newer employees had worked for a major automotive company and had been involved in Lean Program improvement activities associated with production operations. During the design of the new units, several workers asked if they could be allowed to provide input into the design to see if they could make performing their jobs easier, improve efficiency, and improve overall safety of plant operations. The workers' feedback was provided to their immediate crew team lead, but the team lead was focused on ensuring construction goals and forgot to pass on the feedback to the Shift Manager.

The new equipment was installed as a part of the expansion of the facility and Plant Management had evaluated the new equipment and determined that it was so similar to the existing equipment that no additional training was needed for employees, including no new procedures for Operations and Maintenance personnel. Management made this decision largely based on the experience of the workforce and how successful

Chapter eight: Practitioner's guide case study

the company has been in the past. Even though several workers had significant operational experience, the equipment and associated gauges and monitoring systems were highly automated, and the amount of electronic equipment associated with the new systems was far greater than with the existing tanks and processing equipment. The workers expressed concerns about needing additional training and procedures on how each of the ten units operated, along with providing instructions for safe operation of the equipment; however, Management believed they knew what was needed to make the operations successful.

In addition, the Management Team had determined they were going to take a phased approach when starting up each of the new units (each unit would start up in a sequence), allowing one week for equipment testing and readiness preparations before placing the unit into full operation. This approach was decided for startup because the subcontractor responsible for the expansion was significantly behind schedule for construction of the new units, and the phased approach would allow increased production of lacquer while moving toward full production. The phased approach would also allow the company to start expanding business growth sooner into the market for international sales. Installation and startup of the new units were about six weeks behind schedule, and their construction subcontractor was also significantly over budget on the expansion project.

The Plant Manager was concerned about how the production slip would look to his Corporate boss. He didn't fully know how his strategy for phasing-in operations would be perceived, and whether it would ultimately be successful. He was worried about the overall budget and being able to continue to achieve increased profits and revenue with the expansion over budget and behind schedule. He was trying to make a good impression for possible advancement with the D.O. Chemical Company's Corporate Office. The Plant Manager believed that the earlier Spud's Chemical Company could get each new unit online, the sooner their revenue would increase.

The Plant Manager was required to fly back and forth to the Corporate Office almost every week and was therefore not able to spend as much time at his plant as he wanted; however, he had a very competent Deputy Plant Manager and wanted him to be promoted to his position when he received his Corporate promotion. They both shared a common interest in golf and had planned to play in the local Country Club's Member–Guest tournament in two weeks, assuming they would be able to practice that coming weekend. They often discussed business during their time on the golf course. The Plant Manager was becoming frustrated with his Corporate Manager because he was having to do his boss' job and explain to the Corporate President why construction was behind schedule, why increased production goals were not being met, and how they were going to turn it around.

The subcontractor completed construction on Unit No. Nine on Sunday evening August 26, 2018 at about 10:00 pm. The Shift C Work crews reported to work on Monday morning at 6:00 am to start performing the equipment readiness checklist for the newly completed unit. Shift C Work Crews had previously successfully started up operations on five of the eight completed units, and were very proud to be viewed as the most successful crew out of the four; no crew had ever been more successful on the number of unit startups and operation. Over the weekend, the company held their annual picnic, where the Shift C Work crews had been recognized for how quickly they had helped the company reach their production stretch goals. On Monday, everyone felt confident they could start work and didn't need to hold a shift brief, which usually included a discussion on the pre-inspection checklist, any safety issues, lessons learned from previous installations, and any potential operational issues. Shift C Work crews worked all day Monday and made good progress on equipment testing and startup and were able to gain two extra days on the overall project schedule because of how much they were able to accomplish that day. At the end of the Monday, the Deputy Plant Manager had pizzas waiting for the workers to celebrate how much work and progress they had accomplished.

The Shift C Manager came in early on Tuesday morning to talk with the Plant Manager about the progress on Unit No. Nine. The Shift C Manager told the Plant Manager that the work crew had gained two extra days in the project schedule and the Plant Manager encouraged the team to keep that same pace in completing testing and startup for both Units Nos. Nine and Ten. The Shift C Manager talked with the Shift C Team Leads and work crews and relayed the message from the Plant Manager that because of their efficient way of performing startup and testing, the Corporation was considering adopting their approach for all plants owned by the D.O. Chemical Company. The Shift C Work crews continued to work all day Tuesday on equipment testing and readiness startup. One of the Operators told the Shift C Manager that Pump A on Mixing Tank C-1, and two valves upstream of the mixing tank, were not consistently opening and closing like the valves on the other new units. The Shift C Manager told the operator that he appreciated the feedback and would look into the concern the next day. He was in a hurry to leave work because he was in the process of buying a new house and had to leave to sign the closing paperwork before the title company closed.

The Shift C Work crews continued working through the testing and startup process and declared Unit No. Nine ready for full production, Thursday morning, two days ahead of schedule. The Shift C Team Leads were so excited about how they had been able to get the unit started up before any of the other work teams; they were truly the best crew in the plant! The Plant Manager praised the team in front of a group of visiting

Corporate Managers and directed the Shift C Manager to put Unit No. Nine into full production starting Thursday morning, and to begin producing as much lacquer as possible. The Shift C Manager was encouraged by the praise from the Plant Manager and knew this put him in a good position for a promotion and possibly a financial bonus; he was the only Shift Manager to start up five other units and had started up Unit No. Nine ahead of schedule. Shift C Manager was not sure how far behind production they were because Production goals were not published, but he was going to make a positive impact with his team and impress Management.

Shift C Work crews reported to work on Thursday, August 29, 2018 and started up Unit No. Nine and added the unit to the existing process line. The Shift C Operators spent all day filling the storage tanks with the raw materials necessary to operate the unit. In order to get a jump-start on Friday's shift, the Shift C Manager directed the Shift C work crews to prepare the equipment and starting filling mixing tanks. The Shift C work crew had never performed that task before, but they really enjoyed working for the Shift C Manager and wanted to make everyone looked good by staying ahead of schedule and surpassing production goals. They therefore proceeded with the instructions to complete the tasks.

The Shift C Work crews arrived to work on Friday morning, August 30, and started producing lacquer, ahead of schedule for Unit No. Nine. The operators started up all the motors and pumps in the system. At about 10:00 am the Plant Manager requested that half the Shift C work crew assist with the testing and startup of Unit No. Ten; maybe through their help the Plant Manager could gain an additional two days on the project schedule. The other half of the Shift C work crew decided to take a quick break to relax and have an early lunch. At about 10:25 am a loud noise was heard from an area close to Unit No. Nine. Operators from Unit No. Eight went to investigate and upon arrival at Unit No. Nine, found the entire unit was engulfed in flames that reached to the top of the building. The operators quickly called 911 and requested fire response from the local fire department, and also notified the Shift C Manager. The Shift C Manager notified the Plant Manager of the fire and he ran from his office to Unit No. Nine and quickly assisted with evacuating the area. Unit No. Nine continued to burn and subsequent explosions could be heard. The City of Pacer Fire Department arrived; however, they did not know how to fight the fire because no one from Spud's Chemical Company had sent paperwork to the Pacer Fire Department with an updated list of chemicals that were used and stored at the plant. Spud's Chemical Company's Environmental Compliance Manager had left the company a year ago and the company had not yet replaced her. Consequently, the City of Pacer Fire Department had to request assistance from other emergency response agencies. Unit No. Nine, and the area around the unit, continued to burn while additional firefighting and emergency response units arrived on the scene to stabilize the situation; it took 10 hours to extinguish the fire and stabilize the area. Units Nos. Seven, Eight, and Nine all sustained extensive damage, not only to the tanks and process systems, and the electronics for all three units were completely destroyed.

Spud's Chemical Company hired Comprehensive Safety Services (CSS) to investigate the fire and to gain an understanding of what had gone wrong, as well as to learn what could be done to prevent future accidents in the manufacturing of lacquer. The CSS started by interviewing the work crews on Shift C. The work crews were all stunned and could not believe the event had occurred, because the plant had an excellent safety record. However, it was noted that for the past year, safety performance had slipped, but everyone believed it was because of the construction activities associated with the facility expansion. It was recognized that there were areas in the plant where work could be performed in a safer manner, but none of the workers thought something as big as a fire would happen at the plant. When CSS asked the Shift C work crew about whether they felt pressured to get work done, they stated that the plant was behind in their startup schedule and production goals, so they wanted to do everything possible to help. They thought it was acceptable to miss some of the steps in the startup procedures that they deemed redundant because they were the most familiar with the new units. When the CSS asked what the production goals were, none of the workers truly knew since their focus had been on unit startup and testing. Shift C work crews also communicated that none of the goals for the plant were posted as they had been in the past. When asked about when the last time the Plant Manager or any member of the Management Team had spent time in the plant, the response was months ago; he had to travel a lot and spent a lot of time at the Corporate Office. The CSS asked when the last time was that the work crews had attended some type of training, either on the current or the new equipment; they could not remember when they had last received training but pointed out again they were the most experienced team and knew how to operate the equipment. They had recently been praised for their success in equipment testing and startup.

The CSS asked about the layout of the new units, and if it was going to be easier to operate. The workers communicated that they still had to travel about a quarter of a mile to get tools and general supplies for the unit, and that taking readings was probably going to take a little longer than before because of the new layout of equipment. The Safety Manager was questioned about the culture of the company, their safety program, and performance. The Safety Manager indicated that they had an excellent safety record and that safety is the number one priority for the company. The Safety manager stated that they noticed that the Plant Manager was gone a lot and that feedback from the workers did not seem as if it was being addressed. Additionally, the Safety Manager also had to travel

Chapter eight: Practitioner's guide case study 167

to the D.O. Chemical Company Corporate Office frequently to discuss the plant's first-aid cases and recordable injury events. Because of all the travel, and everyone so busy with expansion work, the Safety Manager had stopped providing the weekly safety statistics since feedback was no longer being provided on the safety information provided. Management rarely talked about their safety expectations, but because of the number of experienced workers, they knew they understood how to do their job. The Safety Manager had also recently postponed several safety walkdowns because everyone needed to focus on doing their jobs so that the new equipment could be put into operation sooner.

The CSS representative interviewed individually the Plant Manager, Deputy Plant Manager, Shift C Manager, and direct reports. In addition, several Corporate representatives visited the site while CSS was conducting their review to ensure progress of the review and that Spud's Chemical Company workforce was being managed to D.O. Chemical Company Corporate standards. Also, their job was to assist Plant Management in having a successful interaction with CSS.

During the evaluation, CSS identified that no one had checked the overall design of the new units, or verified that the subcontractor had installed the equipment, as designed by the engineers. The Plant Manager was asked why and he replied that based on the negative project schedule and cost variance, there was not enough time to completely ensure that every aspect of the design was validated prior to installation and startup. The Plant Manager had been creative and decided to validate the design as part of startup testing since he believed that there would be a gain of an extra 42 days in his project schedule by concurrently completing the two tasks.

CSS discovered that in the review of activities leading up to the accident, the Corporation had indeed praised the Plant Manager at a recent Corporate retreat for his ingenuity and ability to improve project performance (the Plant Manager received a small bonus). CSS asked the Plant Manager why he didn't seem to spend much time at the plant. The Plant Manager responded that he had limited time, but felt he had a very competent and people-friendly staff. He did not see the need to spend more time with the workforce and he had Management and Corporate issues that took up most of his time. In addition, the Plant Manager stated that he had a very competent Deputy Manager who was primarily running the plant.

The CSS then conducted walkdowns and in-depth evaluations of the work area and documents associated with equipment testing and startup to determine the cause of the fire. The D.O. Chemical Company had also conducted a Corporate review of the accident, and upon review of their report, the CSS was surprised by the lack of depth, level of detail, questioning attitude, and outcome of the internal investigation. CSS concluded that the fire was caused by faulty wiring installation in Unit No. Nine, coupled with a leaking valve on Mixing Tank C-1. CSS recommended that Spud's Chemical Company develop an operational recovery plan and implement improvements, in both the near- and long-term, to prevent future electrical fires associated with equipment startup. In addition, the CSS provided recommendations that Plant Management evaluate the overall health and culture of the company, and evaluate methods for improving worker performance and engagement, including the relationship between Management and the workforce. Upon discussions with Corporate, the Plant Manager recognized he did not have the time or people to develop such a plan, so he subcontracted the work to CSS to develop the recovery plan and provide recommendations to improve the culture of the company.

chapter nine

The peak performance model applied to Spud's Chemical Company, LLC

9.1 Introduction

This chapter will present the outcome of applying the Peak Performance Model to the scenario provided in Chapter 8. The example will depict how to apply the Peak Performance Model using both qualitative and quantitative analytical methods. As discussed in Chapter 7, the Peak Performance Model integrates Lean, Human Performance Improvement, Operations Excellence, and Organizational Culture in a manner that targets and improves operational and business performance. In the case of Spud's Chemical Company, the checklists, survey questionnaires, and interview questions provided in this book were used to assess the health of the company and identify target areas for improvement. These were then used as the basis for the development of an improvement plan. If appropriately designed, the improvement plan should drive Spud's Chemical Company closer to achieving Peak Performance.

9.2 Qualitative evaluation of Spud's Chemical Company

Spud's Chemical Company had a significant incident when initiating operations on Unit No. Nine. The work crew did not utilize a well-thought-out or documented process, such as a procedure or a checklist, for starting up the unit. One of the employees noted that Pump A on Mixing Tank C-1, and associated valves, were not operating like the others and mentioned it to their Supervisor. However, the Supervisor did not follow up on the issue and the unit failed, causing an explosion and fire. Spud's Chemical Company, with assistance from Corporate (D.O. Chemical Company), brought in resources and conducted a review to develop corrective actions and ensure a similar event would not occur again at Spud's Chemical Company. After completing the review, the company and Corporation recognized that Spud's Chemical Company had more significant challenges than just recovering from the incident. Spud's Chemical Company hired

Comprehensive Safety Services to perform an independent evaluation of the company and to develop an operational improvement plan that would result in demonstrating improved operational performance while driving sustainability of operations and company success. Comprehensive Safety Services applied the Peak Performance Model and performed both a qualitative and quantitative evaluation of the company, and the incident, to support development of an effective operational improvement plan.

9.2.1 Lean program activities

Comprehensive Safety Services reviewed policies and procedures, conducted multiple interviews, and observed employees performing operational tasks to gain an understanding of how Spud's Chemical Company operated, and to learn more about the Management Team and whether or not Lean program methodologies were implemented. Below is an excerpt from an interview with the Plant Manager at Spud's Chemical Company. Interview responses directly correlated to the Lean Checklist contained in Chapter 2. Following are notes from the assessor detailing discussions with employees, as well as observations noted, while walking through the plant.

1. *Question:* Could you please tell Comprehensive Safety Services members your name, background, and how long you have been at Spud's Chemical Company?
   Response: My name is Bill Black; I have been the Plant/Facility Manager at the Spud's Chemical Company for 2.5 years. Before arriving at this facility, I was the Operations Manager at the Ruby Rabbit Company for 7 years. I have a total of 20 years' experience in operations.

2. *Question*: Does Spud's Chemical Company have a strategic plan for implementing a Lean Process?
   Response: The plant does not need a strategic plan, and the facility performed well previously before the upgrades began.
   Assessor Notes – When touring the plant, no signs of a strategic plan were visible nor did any of the employees we talked with know of any strategic plan. In fact, most employees were not familiar with Lean principles. Comprehensive Safety Services did interview a couple of newer employees that understood the principles, but it was from a previous employer.

3. *Question:* Do you and your managers walk down areas with a "Lean Lens" looking for potential areas of waste?
   Response: No, we do not need to spend our time looking for waste because the facility operates well and I don't believe we have any additional waste to identify.

Chapter nine: Spud's Chemical Company, LLC 171

Assessor Notes – The employees interviewed stated they seldom saw Management in the plant. The Management Team never discussed looking for process improvement opportunities or waste; Management just told them to work faster so they could meet the production goals.

4. *Question:* Has training been provided to Senior Management on Lean principles?
   Response: No, I am not aware of any training on Lean principles. We had that "Lean" training at the Ruby Rabbit Company and I thought it was a waste of my time. *Assessor Notes* – The Comprehensive Safety Services asked if the employees were given any training on Lean Principles, and the response was not since he had been at the facility.

5. *Question:* Once waste has been identified do your managers develop a plan to identify, measure, analyze, improve, and sustain improvement measures? Have you asked your Management Team if their employees have provided any recommendations?
   Response: I am not aware of anyone that has identified areas of waste, so no need to develop any plans! I am sure if the employees had recommendations, I would have been told.
   Assessor Notes – Employees stated they tried to give input on recommendations to improve, but never heard anything back from the Management Team, so they just stopped looking and telling their Management.

6. *Question:* Do you and your Management Team review progress toward actions to reduce the waste?
   Response: We don't have any waste – so no need to track progress.

7. *Question*: Does the Management Team value the Lean or other improvement processes and encourage the employees to become engaged?
   Response: I don't have any value for that stuff, and I do not discuss Lean with my staff, we have more important issues to deal with.
   Assessor Notes – Interviewed employees with Lean experience and they expressed that they had tried to provide recommendations on how to make the design better and make their jobs easier; however, they never received a response back from the Management Team.

8. *Question:* Are metrics/goals reviewed on a routine basis by the Management Team?
   Response: I know my production goals and how to meet them, so we don't need to review them, just meet them.
   Assessor Notes – Employees commented on not really knowing the goals because they never see them.

9. *Question:* Do your Managers discuss opportunities to utilize Lean concepts?
   Response: We don't have time to discuss our production goals; we just need to meet them.

10. *Question:* Are your Managers driving for daily improvement, i.e., looking for incremental improvements? How does your team go about identifying these improvements?
    Response: We are always looking for ways to meet our production goals, so I guess you could say we are always looking to improve. I am not sure how they look for improvements, I have never thought to ask.

11. *Question:* Do workers actively participate in a suggestions/ideas program?
    Response: We have been operating the same way for a long time, we have the plant operating the way we want, I don't believe the workforce has any ideas on how to do it better, and if they did, I would have heard them.

12. *Question:* Are metrics visible to the workforce?
    Response: I do not see any need to post stuff on the walls. We know what we need to do.
    Assessor Notes – The workers commented about the lack of visible metrics.

13. *Question:* Are employees visibly engaged in reducing waste in the process?
    Response: Of course, our employees are engaged, they come to work every day and do their jobs. We don't have a lot of waste as we operate efficiently.

14. *Question:* Are results from the Lean events published to the workforce?
    Response: The workforce knows our process, so I don't see any reason to post stuff.

15. *Question:* Do employees volunteer to participate in Lean activities to support elimination of waste?
    Response: Our facilities are efficiently operated, we don't have a lot of waste, therefore we don't need to conduct Lean activities.

16. *Question:* Do workers seek out the Management Team to discuss opportunities to improve?
    Response: If my workers have something to say, they will tell me; I always have an open door and believe I am very approachable.

17. *Question:* Have the workers been trained in Lean concepts and tools or other improvement processes?

Response: I don't think we need to train our employees in Lean, they know how to do their jobs effectively.

18. *Question:* Are various tools utilized to reduce waste, i.e., fish-bone diagrams, etc.?
    Response: I think I remember hearing about some Lean tools at the Ruby Rabbit Company, but I have never used them.

19. *Question:* Is rework tracked and trended?
    Response: My folks do a great job, we don't have a lot of required rework.

20. *Question:* Is standard work utilized? How you do verify standard work is used?
    Response: Yes, we have been using the same equipment for years along with the same employees. We do it the same way every day. There is only one way to operate the equipment, so I don't need to verify we are using standard work.

21. *Question:* Are the 5S principles utilized in the work area?
    Response: I don't know what 5S is, so I don't think we are using it.

22. *Question:* In what areas has Lean been implemented (procedures, operations, work control)? Is it evident?
    Response: I don't believe in that Lean stuff, so we have not implemented it anywhere, we are just focused on production goals.

23. *Question:* Has a time–motion study been conducted to highlight areas where employees may have excessive waiting periods?
    Response: No study has been completed – we don't need a study to tell if our folks are busy or not. We know they are busy all day long.

24. *Question:* Is work planned utilizing Lean concepts?
    Response: We have done our jobs hundreds of time, we don't need to plan our work, we know the activities we are required to complete.

25. *Question:* Do your Managers encourage workers to make recommendations/suggestions?
    Response: If employees have suggestions, I have never heard of them. This workforce just comes to work, does their jobs, and then goes home.
    Assessor Notes – Employees provided comments and wanted to provide input into the new design. The Management Team is doing just the opposite; they are discouraging the workforce by not providing any response or follow-up.

26. *Question:* Are employees fully utilized as an important resource?
    Response: My employees work very hard, they go home every day tired from a good day's work. Of course, our employees are important to us.

27. *Question:* Are the opportunities for Lean, or some other improvement process, identified, and communicated to the workforce?
Response: We communicated with the workforce on important issues, like where we are on meeting the production milestones.

28. *Question:* Are visual management tools utilized in the facility (production metrics, improvement actions, etc.)?
Response: I don't believe in posting a lot of stuff on the walls, it distracts the workers. We just need to stay focused on getting the job done!

29. *Question:* Does the organization have a visible mission statement that employees know and identify with?
Response: We all have one mission – meet the production goals. I am sure my employees know that.

30. *Question:* Is the plant meeting its goals? If not, why not?
Response: We were struggling to meet our production goals even before the accident. I am really not sure why, it could be the new employees are just not working as hard as the old ones, I really don't know why, never really thought about why, I have just been focused on meeting the goals. You know if we don't meet our goals, the Corporate Office is calling and yelling at us, so I try to avoid those kinds of calls.

9.2.1.1 Lean summary

Evaluation of Spud's Chemical Company indicated that the leadership team was not Lean-thinking managers. The company had not implemented a Lean process and had very little interest in Lean thinking and process improvements. Comprehensive Safety Services identified several weaknesses that Spud's Chemical Company should address to make existing work processes more efficient and client-focused. Below, are the key company process weaknesses as related to the following four shared attributes: leadership, employee engagement, organizational factors, and work planning and execution.

9.2.1.1.1 Leadership
- The Management Team did not set or communicate expectations to the workforce
- The Management Team spent limited time in the facility interacting with the workforce. This led to the employees not feeling valued or appreciated
- The Management Team did not actively show support for the Lean or any other type of improvement process

9.2.1.1.2 Employee engagement
- The employees tried to provide recommendations to make the facility better, but the Management Team did not take actions on any feedback or recommendations
- The workforce was unable to gauge progress and drive continuous improvement

9.2.1.1.3 Organizational factors
- The Management Team did not clearly communicate goals to the workforce
- Visual management tools were not used to status objectives or goals

9.2.1.1.4 Work planning and execution
- The Management Team did not frequently interact with the workforce
- Lean principles were not incorporated into work planning and execution
- Standard work processes were not utilized by the workforce, thereby increasing the potential for errors
- Time–motion or other studies had not been conducted to highlight opportunities to improve workflow

9.2.2 Human Performance Improvement (HPI)

Comprehensive Safety Services conducted multiple interviews, reviewed several procedures and policies, and observed the performance of operational tasks to gain an understanding of how Spud's Chemical Company operated and to learn more about the Management Team in relation to an HPI program. Here is an excerpt of an interview with the Plant Manager at Spud's Chemical Company, and assessor's notes from interactions with employees, and observation of work.

1. *Question:* Could you please tell Comprehensive Safety Services members your name, background, and how long you have been at Spud's Chemical Company?
   Response: The worker said his name was Bill Black. He has been the Plant Manager at Spud's Chemical Company for two and a half years. Before arriving at this facility, he was the Operations Manager at the Ruby Rabbit Company for seven years. He as a total of 30 years' experience in operations.

2. *Question:* Does Spud's Chemical Company have any visible management support for the Human Performance process? Do the Corporate offices require any focus on Human Performance?

Response: Bill said he did not really understand what Human Performance was, so he guessed there was not any support. When asked if Corporate required any focus on Human Performance Improvement, he said he had heard about Human Performance while at the Corporate Offices but it was not required to be instituted at his plant.

Assessor Notes – When touring the plant, no signs of Human Performance Improvement implementation was visible to the team. In fact, most employees were not familiar with the principles. The Comprehensive Safety Services did interview a couple of newer employees that understood the principles, but it was from a previous employer.

3. *Question:* Does the plant have a strategic plan to reinforce the implementation of Human Performance Improvement principles?
   Response: Bill stated that he was not familiar with any strategic plan for his plant or even at the Corporate Offices.

4. *Question:* Was Human Performance training provided to the workforce both initial and refresher training? Did the leadership team ever receive Human Performance training?
   Response: No training was provided to the newer employees and it did not appear that Human Performance training was ever provided to any of the employees. Bill stated he was not aware of the leadership team ever having any type of Human Performance training at his facility; however, he thought that the Corporate Office may have provided training in the past.

5. *Question:* Are Human Performance metrics maintained and trends evaluated to make midcourse corrections?
   Response: No metrics were discussed or visible to the workforce. Bill stated that they did not have any metrics at all, nor did they need them.

6. *Question:* Any evidence that Human Performance is included as part of work planning?
   Response: Bill stated that they did not really need Human Performance as a part of the work planning process, his team was very experienced and already knew what to do.

7. *Question:* Any evidence that Human Performance is included as a part of post-job reviews?
   Response: Bill stated that they did not conduct post-job reviews, there was no time; they had production goals to meet. They really did not even have time for staff meetings.

8. *Question:* Is feedback solicited and acted upon?
   Response: Bill stated that if his team gave him any feedback, he would act upon it, but he has never been given any feedback.
   Assessor Notes – Several of the newer employees had recommendations on the design to improve the process, but Management never requested their input.

9. *Question:* Are metrics and trends discussed with the workers to support improvement initiatives?
   Response: Bill stated that his team did not need to be distracted with pictures and charts on the wall, they just needed to keep their heads down and work.
   Assessor Notes – No metrics were visible or used anywhere in the plant.

10. *Question:* Was the Management Team visible in the plant coaching the workers to reinforce expectations and obtain feedback?
    Response: Bill stated he had very little time to spend in the plant, he spent most of his time in the Corporate Office to keep the bosses happy. However, he knew his Management Team spent the majority of their time on the plant floor.
    Assessor Notes – The employees stated that they seldom saw any Management in the plant. If they saw Management, it generally meant that someone had screwed up and the whole group was about to hear what happened!

11. *Question:* Did the Management Team discuss error-likely situations with the workers?
    Response: Bill stated that his team was competent and did not usually make errors. If an error is made, it is generally one of the newer workers.
    Assessor Notes – When workers were asked about discussions around error-likely situations, most did not understand the concept. Some of the newer workers stated that they did not ever discuss any Human Performance principles or tools at the plant. The Plant Manager missed a good opportunity to discuss error likely situations when Shift C began the startup of Unit No. Nine.

12. *Questions:* Were lessons learned discussed with the workers?
    Response: Bill stated that his team was really good; they did not need or have time to discuss that stuff.
    Assessor Notes – The Management Team missed a good opportunity to discuss lessons learned from the previous startup before checking out the equipment and starting up Unit No. Nine.

13. *Question:* Were critical steps reviewed with the shift workers prior to starting a job or a specific task?
    Response: Bill stated his team did not need those types of conversations or be involved in a lot of preplanning activities because they knew what needed to get done and they just did it.
    Assessor Notes – Multiple opportunities were missed to discuss critical steps, at the start of the activity, after the equipment was checked out, and prior to starting up the unit.

14. *Question:* Do workers discuss critical steps among themselves looking for lessons learned from coworkers?
    Response: Bill stated he was not aware of his team having these types of discussions. *Assessor Notes* – Multiple opportunities were missed to discuss critical steps, especially with several new activities that crews were conducting that others had done previously.

15. *Question:* Were self-assessments conducted to review the implementation of Human Performance tools?
    Response: Bill stated that they did not have time for assessing their performance; they had production goals to meet. In order to complete an assessment, someone would not be working on production and that would not go over well with Corporate since they expect them to meet their production goals, not spend time doing unproductive work.

16. *Question:* Did the organization display a questioning attitude, or were they interested in learning and getting better, improving efficiency and overall production?
    Response: Bill stated they did ask questions when production goals were not met.
    Corporate would also ask why production goals were not being met.

17. *Question:* Does the organization conduct benchmarking activities with other companies to identify best practices?
    Response: Bill stated that they did not conduct benchmarking activities. That would be something that the Corporate Office would do. However, he was not familiar with any benchmarking activities.

18. *Question:* Does the organization have a process in place to observe work activities and behaviors in the plant to gain a better understanding of how work is actually accomplished?
    Response: Bill stated that they did not have time to observe work. Most of his workers were experienced and knew what they were supposed to do.

19. *Question:* Does the organization utilize independent oversight as a way to uncover potential blind spots?

Chapter nine: Spud's Chemical Company, LLC   179

Response: Bill stated that he was not aware of any group conducting any independent oversight activities in the plant. His said if that happened it would be coordinated by the Corporate Office.

20. *Question:* Does the organization utilize a change management process before implementing significant changes (such as large organizational changes, major changes in operations, etc.) to limit the overall impact to the workers?
Response: Bill stated that he was not aware of a change management process. He stated that the company made very few changes in policies and work practices.
Assessor Notes – An opportunity was missed to develop and implement a change management plan before designing and starting up 10 new units.

21. *Question:* Does the organization report errors and near-misses to improve performance?
Response: Bill stated that he did not believe his team made many errors, so no need to report the few they might make. No one has time to keep a list anyway.

22. *Question:* Do employees feel free to bring up issues to the Management Team?
Response: Bill stated that he believed his team would bring up issues if they had any.

23. *Question:* Do employees believe that if they bring up an issue that the Management Team would act to resolve?
Response: Bill stated that he would resolve their issue if it was brought to his attention.

24. *Question:* Are task previews conducted?
Response: Bill stated that his team was very experienced and knew their job, so why waste time conducting a task preview, they had done these jobs thousands of times.

25. *Question:* Are pre-job briefings conducted? If so when?
Response: Bill stated they might do a pre-job brief, but not often. His team did not need it. When asked when a pre-job briefing would occur, he replied that there was not a set policy or timeframe when they should be performed.

26. *Question:* Are critical steps discussed during the pre-job brief?
Response: Bill stated that all the steps were critical – so no need to talk about all of the procedure. Besides, his team was very competent; they would only do these things for an inexperienced group of workers.

27. *Question:* Are error-likely situations identified/discussed?
    Response: Bill stated his team does not make errors, so no need to discuss.

28. *Question:* Do workers follow written procedures?
    Response: Bill said his team has procedures to follow. However, they are very experienced and therefore really do not need them. He believes a few of the newer workers must use them.
    Assessor Notes – The workers requested that the procedures and training be revised for the upgrades, especially the equipment, associated gauges, and the monitoring systems because they were highly automated. All were very different from the old production lines.

29. *Question:* Do workers utilize good communications techniques during the performance of tasks?
    Response: Bill stated his team communicated well with each other, they all get along well.
    Assessor Notes – Overall communications were not clear during the accident; the employees were not sure what needed to be done or who was in charge during the event.

30. *Question:* Are issues openly discussed and followed up on?
    Response: Bill stated he always followed up on concerns, but he seldom had any.
    Assessor Notes – During startup of Unit No. Nine, a worker brought up the issue with Pump A on Mixing Tank C-1 and the two valves upstream of the mixing tank that were not consistently opening and closing like valves on the other new units. However, no follow-up occurred, and the worker did not voice the concern again.

31. *Question:* Are post-job reviews held after tasks are completed to identify issues and add lessons learned into the work package/checklist before the task is performed again?
    Response: Bill stated no need to have post-job reviews. They did not need them, his team knew what was needed. However, reviews were considered and infrequently performed.
    Assessor Notes – At the completion of each new unit was an opportunity to conduct a post-job review to discuss lessons learned and make changes to the checklist prior to starting up the next unit.

32. *Question:* When workers are required to perform work on similar or lookalike components, does the team use flagging techniques to ensure the proper piece of equipment is worked on?
    Response: Bill stated he was not aware of any time his team would have used that technique.

9.2.2.1 Human Performance Improvement (HPI) summary

The evaluation showed that the company had not implemented an HPI program and had very little interest in doing so. Using the assessment tool described in Chapter 3, Comprehensive Safety Services identified several weaknesses that Spud's Chemical Company should address to reduce the potential for errors and become more efficient. Following is a summary of those weakness organized into one of four shared attributes for HPI: leadership, employee engagement, organizational factors, and work planning and execution.

9.2.2.1.1 Leadership
- The Management Team does not discuss error precursors
- The Management Team does not conduct task previews
- The Management Team does not discuss critical steps when performing tasks

9.2.2.1.2 Employee engagement
- Since the Management Team had limited engagement with the workforce, Management did not understand the problem areas and challenges the workforce routinely encountered
- The Management Team did not seek feedback routinely from the workforce
- Since the Management Team had limited engagement with the workforce, Management did not understand the problem areas and challenges the workforce routinely encountered
- The Management Team did not seek feedback routinely from the workforce

9.2.2.1.3 Organizational factors
- It is difficult to improve performance without clear communications of expectations and goals with the workforce
- The Management Team did not create a culture of continuous improvement
- Because of the limited time spent with the workforce, the ability to build trust was diminished
- The Management Team failed to openly address issues

9.2.2.1.4 Work planning and execution
- The Management Team did not evaluate completed tasks to understand what went well when performing tasks and what could be improved
- There was no visible indication of Human Performance tools being used during the conduct of work
- Limited effort was spent reducing the potential for errors

9.2.3 Operations excellence

The Comprehensive Safety Services conducted both document reviews and worker interviews to gain an understanding of how Spud's Chemical Company operated, and to learn more about the Management Team in relation to an Operations Excellence program. The following section provides a summary of the Operations Excellence analysis for Spud's Chemical Company.

9.2.3.1 Document reviews

Comprehensive Safety Services reviewed documents such as policies, procedures, worksheets, checklists, and guides to determine whether the documentation reflected aspects of an Operations Excellence Program. It was apparent to Comprehensive Safety Services that formally identifying those documents of Spud's Chemical Company that supported a more structured, formal approach to operations would be of benefit, and would contribute to a safer, more productive workplace. Comprehensive Safety Services determined that Spud's Chemical Company was void of key interfaces needed to sustain excellence in operations, and that the company could benefit from an Operations Excellence program.

9.2.3.2 Interview questions and responses

Following is an excerpt from the interview with, and responses from, Spud's Chemical Company Deputy Plant Manager. Questions are from the Operations Excellence Checklist contained in Chapter 5.

1. *Question:* Could you please tell the Comprehensive Safety Services members your name, background, and how long you have been at Spud's Chemical Company?
   Response: My name is Ted Chloe; I am the Deputy Plant Manager at Spud's Chemical Company. I have worked for Spud's Chemical Company for approximately 25 years (including time at the former company).

2. *Question:* Has Management and Supervision communicated their values and commitment to the organization?
   Response: Every once in a while we hold an employee gathering where the company values are discussed, but not often. The last big company meeting was at the company picnic and we got to talk there. I try not to distract my people from their job.

3. *Question:* Do Management and Supervision spend more than 20% of their time in the plant interacting with the workforce?
   Response: I have no idea. I am too busy trying to meet our production goals and dealing with Human Resources. They know they are

Chapter nine: Spud's Chemical Company, LLC   183

supposed to be in the plant watching the workers so that we can meet our production goals.

4. *Question:* Does Management routinely conduct organizational meetings to reinforce their expectations for performance?
   Response: We hold meetings when we need to. We have to be careful not to unnecessarily pull people from their main job.

5. *Question:* Do Supervisors provide feedback to workers as to how well they believe their team is performing?
   Response: We let the workers know what a great job they do every day. If you were to go ask them, they would tell you Leadership has a great relationship with the workforce.
   Assessor Notes – Comprehensive Safety Services performed both a culture and a trust survey and found there was a lack of trust, and the culture of the organization was marginal.

6. *Question:* Does Management and Supervision verbally and through their actions hold themselves accountable for the performance of the organization?
   Response: If a mistake is made, then whoever makes it is held accountable.

7. *Question:* Does the Management of the organization embrace self-reflection and learning as a fundamental tenet of the company?
   Response: We are always trying to learn and get better. The Corporate Office wants us to try and always do better. We need to so that we can increase our productivity – of course we do. I can't believe you are even asking me that question.

8. *Question:* Does Management include representatives of the workforce in their strategic planning for the company?
   Response: We don't need the workforce to drive planning for the company, we need the workers to perform their jobs and let managers do their job.

9. *Question:* Do workers understand their role and how it contributes to the overall success of the company?
   Response: Everybody knows what they are supposed to be doing. We provide a full day of training to workers on how to operate equipment and how the company works.

10. *Question:* Is worker feedback considered in the acquisition of people, equipment, and in the development of company policies and procedures?

Response: Not to my knowledge. We generally listen to what they have to say, but it is Management's job to acquire people and equipment. You have to be careful because there is a specific profit margin, from Corporate, that Spud's Chemical Company is required to meet.

11. *Question:* Are workers involved in the development of organizational training and is the feedback considered in the development of retraining?
Response: Our training classes are very good and we have never had a complaint about our training programs. As always, we listen to the workers when they have feedback, but we have to be careful on making sure we do not overspend our budget on unnecessary training; there is a careful balance between providing what is necessary versus training that is wasteful and adds no real value to our company.

12. *Question:* Are workers alert and attentive to control-panel indicators, alarms, and the work environment?
Response: There are times when it can be slow, but as a general rule the workers are alert and know how to perform their jobs. Now that we have switched to a 24-hour-shift, seven-days-a-week operation, we may want to look at being more receptive to worker feedback. However, if something went wrong we would immediately be notified.
Assessor Notes – Workers complained about having to work so much overtime that at times they found it difficult to concentrate on the job. They were afraid to raise the issue for fear of losing their job because over the past year, several workers tried to communicate with Management on how to better perform their job, but soon after the workers were either reassigned to a "more important task" or let go. They were viewed as "hard to work with," and Management didn't believe they were working to help the plant be successful.

13. *Question:* Do workers monitor control-panel indicators frequently and take prompt action to determine the cause of, and correct, abnormalities?
Response: The workers should always be watching their operational control panels; that is their job. When a problem does happen, the workers know that they are to immediately notify their Supervisor.

14. *Question:* Are communications clear and concise? When communicating verbally, is repeat-back used as well?
Response: Every time, everyone knows how to talk to each other. We don't need to repeat back information; most people hear it the first time.

Chapter nine: Spud's Chemical Company, LLC   185

15. *Question:* Do oncoming and off-going workers walkdown their work environment together?
    Response: It is a rule that the workers are supposed to walkdown their environment.
    That is part of the Supervisor's job, as well as, the worker's job to make sure that happens.
    Assessor Notes – Several of the experienced workers communicated they use to walkdown their areas, but after the company was bought by the D.O. Company, Supervision wanted them to start work at assigned locations within 10 minutes of clocking in. The company even installed a new system for clocking-in workers and remotely monitoring their time on the job.

16. *Question:* Is worker feedback solicited when investigating an abnormal event?
    Response: I am not sure of what you consider an abnormal event. When something goes wrong, we make sure it is fixed. That is Management's job, to ensure issues are corrected.

17. *Question:* Is the process for planning and executing work documented and understood by Management, Supervisors, and workers?
    Response: Yes. Everyone knows how to do work and we really don't need to do much planning because it is a manufacturing plant and we have procedures. It is impressive, most of the workers don't use the procedures and know how to operate the equipment without any problems. We have been able to gain a 5% efficiency since instituting the requirement to start working within 10 minutes of clocking in.

18. *Question:* Is a graded approach applied to the planning and execution of work?
    If so, do Supervisors and workers understand their roles and responsibilities?
    Response: We don't really think about planning work; we just follow our processes and procedures for performing work. We have all been trained and know what we are supposed to do.

19. *Question*: Are changes in configuration and status of processing systems and equipment communicated to workers and provided as a part of shift change?
    Response: Yes. We talk about how the plant is operating all of the time. If something changes, workers know they are to tell their Supervisor.
    Assessor Notes – Workers provided feedback that since the requirement to start work so quickly after clocking in, they rarely have time to coordinate an area walkdown before starting to work.

20. *Question:* Are equipment deficiencies identified by workers?
    Response: Yes – if something doesn't work, they let supervision know.
    Assessor Notes – Comprehensive Safety Services noticed a significant maintenance backlog when reviewing documentation. When Comprehensive Safety Services inquired about the backlog, the Maintenance Manager said they were working as hard as they could, but since the company has created a 24-hour shift, the maintenance backlog seemed to have grown larger and the company had a Corporate challenge to decrease maintenance evolutions by 10%.

21. *Question:* Is equipment tested following maintenance to demonstrate that it is capable of performing its intended function?
    Response: We just leave that to the Maintenance group. I don't tell them how to do their job and they don't tell me how to do mine.

22. *Question:* Is information transferred at shift turnover accurate and does it provide the oncoming shift with information on plant status and actions needed to maintain the plant in a safe condition and to continue operations within an acceptable operating envelope?
    Response: When there is a change in shift, the Supervisors are supposed to be reviewing paperwork to ensure they are aware of any problems or issues.
    Assessor Notes – When this same question was put to the workers they communicated that it is pretty rare that they all meet and discuss what happened on the shift before.

23. *Question:* Are procedure changes documented and approved?
    Response: We don't really need to document our procedure changes; we just red-line out the old steps and write in new ones. When the procedures get to the point where it gets confusing or hard to read, then we take the time to change the procedure. We are here to make lacquer, not to generate paperwork.

24. *Question:* Is worker feedback actively solicited and a process established to incorporate feedback into the work planning process?
    Response: When a worker comes to me and tells me they have an idea to improve how we perform work, then I listen. I try and encourage workers to let us know of improvements; it will help us become more profitable. However, the workers know they need to perform their job.

25. *Question:* Do workers feel they can trust their Supervisors, Mid-Level Managers, and Company leadership team?
    Response: Yes – absolutely. Everyone knows we are the best company to work for in North Carolina, and certainly in the Pacer area. The

workers love the Management Team here; you saw that at the company picnic.
Assessor Notes – A trust survey administered by Comprehensive Safety Services found a low level of trust in the organization and subsequent Management Team.

26. *Question:* Do workers believe Management is honest when communicating with them?
Response: Of course, we are honest, for you to ask that question shows you do not understand our Company. We are open and honest in all aspects of running the company.

27. *Question:* Do workers believe Management is effectively communicating, and reinforcing through their actions, core values of the company?
Response: I personally believe the Company spends too much time advertising who we are. I really have never understood what a core value is and what they mean. We manage the company in an honest manner and I really don't appreciate you questioning our values.

28. *Question:* Does Management promote honesty and trust as a core business practice?
Response: Always. Everyone knows we are honest and trustworthy. We are a large employer in the Pacer area and give a lot of money away to the community.

29. *Question:* Does Management demonstrate concern for employees through their actions, policies, and the mechanisms used to engage workers?
Response: Yes. The Company is very caring. For example, we just had a Company picnic several weeks ago and workers told us how much they appreciated the picnic and their prizes. We even gave away money as prizes; I don't know how you could be more caring than giving away money.

9.2.3.3 *Operations excellence summary*

Evaluation of the Spud Chemical Company indicated the Company had not fully implemented an Operations Excellence program; and therefore, were not optimizing business processes and practices. Based on interviews and document reviews, Comprehensive Safety Services identified the following weaknesses that should be addressed to make existing work processes more efficient and client focused. Below are the key company process weaknesses organized by four shared attributes depicting an Operations Excellence Program: leadership, employee engagement, organizational factors, and work planning and execution.

9.2.3.3.1 Leadership
- Management does not include workers in daily planning or strategic planning sessions
- Management does not appear to be open to worker feedback
- Management and Supervisors are significantly driven to manage the company keeping production costs low and profits high
- Management spends minimal time in the plant
- Company values and commitments are not routinely communicated

9.2.3.3.2 Employee engagement
- Minimal training is provided, unless a special request is made
- Worker feedback is not solicited and possibly not value
- No formal policies exist for ensuring consistent implementation of overtime, communications, and operational analysis of information
- No formal documents exist that establish minimum organizational expectations associated with formality in operations

9.2.3.3.3 Organizational factors
- There is a low trust level that exists between workers and the Management Team
- Management does not understand the purpose of Corporate values and commitments
- Management does not exhibit an awareness of how workers are managed can directly influence productivity
- Management is focused on increasing productivity goals versus understanding how to improve operational performance to achieve an increase in revenue

9.2.3.3.4 Work planning and execution
- The work planning process is not clearly defined
- Management does not recognize that a robust equipment maintenance program is a top priority when desiring to achieve peak performance
- There is a lack of recognition by Management that workers can help improve the way work is performed
- Adequate time is not planned for personnel to share information during shift change

9.2.4 Organizational culture

Comprehensive Safety Services evaluated the culture of the Spud Chemical Company using several methods. These methods included document reviews, surveys, focus group discussions, and individual interviews. In addition, survey instruments were administered to workers to gauge their perception of the culture along with their perception of key aspects of the organization. The surveys were specifically designed to measure overall culture, employee engagement, and organizational trust.

9.2.4.1 Document review

A review of the guiding documents, depicting how Spud's Chemical Company operates and performs their business functions, including procedures, policies, and charters was conducted. These documents were reviewed to determine if organizational expectations were established and communicated. Results of the document review indicated the following:

- The organization had several committees in place; however, none of the committees had written charters that would guide how the committees would function, support the mission, or enhance organizational practices or vision. One example noted was the company's safety committee. Lacking a written charter, Comprehensive Safety Services observed that the meetings were unorganized and appeared to be used by employees as a forum to complain about leadership and discuss safety issues that had occurred. The meeting often did not yield solutions that could be communicated to the workforce or the rest of the company. In addition, the committee lacked Senior Management sponsorship to promote issue resolution and serve as a bridge to remove barriers as they occur. Lacking Management sponsorship and involvement, members of the committee stated they felt helpless in their ability to make a difference in championing and leading change
- Many of the procedures reviewed (20 out of 28) by Comprehensive Safety Services had not been reviewed and updated since the D.O. Chemical Company purchased Spud's Chemical Company

9.2.4.2 Employee engagement survey

Comprehensive Safety Services sent out an employee engagement survey to all employees at Spud's Chemical Company, with a return rate of 30%. The survey contained multiple questions to engage employee perceptions of how involved they are with organizational matters. Examples of the questions that were included in the survey can be found in Chapter 6. Table 9.1 shows the employee engagement survey scale used.

Table 9.2 lists several example questions that were used for the employee engagement survey. Comprehensive Safety Services reviewed the mean for each survey question and identified improvements that were needed in organizational leadership, interactions with the workforce, and improvement in employee relationships. The survey mean was 2.36 for the entire company.

Table 9.1 Employee Engagement Survey Scale

Scale	Corresponding Meaning
1	Always
2	Sometimes
3	Usually
4	Seldom
5	Never

Table 9.2 Employee Engagement Questions of Focus

Question	Mean
2. My opinion is often sought by Management	2.0
7. I have the opportunity often to participate in providing input on decisions affecting me and my team	3.0
10. My Management team inspires me	2.5
18. I would happily recommend my organization as a good place to work	2.3
22. Management encourages worker involvement in organizational activities	2.1

9.2.4.3 Trust survey

Comprehensive Safety Services sent a trust survey to all employees at Spud's Chemical Company, with a return rate of 35%. The survey contained multiple questions. Examples of the types of question can be found in Chapter 6. Table 9.3 shows the trust survey on a scale from 1 to 5. Comprehensive Safety Services reviewed the survey and found a trust-level overall score of 2.9, which shows some level of trust in the Management Team. Table 9.4 lists key questions used in the trust survey.

9.2.4.4 Organizational culture survey

Comprehensive Safety Services sent out an organizational culture survey to all employees at Spud's Chemical Company, 40% of which were

Table 9.3 Organizational Trust Scale

Score	Trust Level
1	Minimal
2	Little
3	Some
4	Great
5	Very Great

Chapter nine: Spud's Chemical Company, LLC

Table 9.4 Trust Questions of Focus

Questions	Mean
1. Senior Management in my company communicates information completely and frequently	2.0
2. The Managers in my organization keep promises	1.9
6. Clear, concise communication is seen throughout the organization	1.8
8. Communication flows in all directions in my company to ensure that all workers are kept informed	2.6
9. Disagreements and issues are addressed in a timely manner	2.5
10. My Managers support a work–life balance for workers	2.7
11. Managers in my organization openly admit mistakes when they occur	2.2
13. Management always communicates openly and honestly	2.4
16. Information is freely and willingly shared by organization members	2.1
19. Managers in my organization demonstrate good leadership and management skills when conducting business and making decisions	2.3

Table 9.5 Organization Culture System

Organic (adaptive)						Mechanistic (change averse)	
1	2	3	4	5	6	7	8
1st quadrant		2nd quadrant		3rd quadrant		4th quadrant	

Table 9.6 Organizational Culture Questions of Focus

Question	Mean
3. I believe that my Management values my opinion	6.1
6. The mission and values of my organization are posted for employees to view	7.1
7. In my organization, Management celebrates the successes of employees at every level	6.3
9. Management is responsive to suggestions from employees	7.0
17. Employees have the right training and skills to perform their jobs	6
20. Morale is high in my organization	6
23. Employees speak highly of my organization	7
24. Roles and responsibilities in my organization are clearly defined and understood	6
26. My Supervisor is a positive role model	6

returned. The survey contained multiple questions, which can be found in Chapter 6. Table 9.5 shows the organizational survey on a scale from 1 to 8. Comprehensive Safety Services reviewed the survey and found an overall organizational culture score of 6.08. Comprehensive Safety Services recognized that it would not be feasible to target all areas for improvement; questions identified in Table 9.6 had a mean in the upper third and fourth quadrants and were used to target improvements in the culture of the organization.

9.2.4.5   Organization culture focus-group interview

In order to further understand the survey results, Comprehensive Safety Services facilitated focus-group interviews with six groups, each consisting of six workers. The groups were asked multiple questions with their responses summarized as follows.

1. *Question:* What words would you use to describe the culture of your organization?
   Response: Not very engaging for employees or managers, closed to accepting outsiders such as new employees, stuck in the past when it comes to innovative ideas.

2. *Question:* How does communication flow in your organization?
   Response: Flows well between Supervisors and direct reports, oftentimes the rumor mill is used because the information provided is incomplete or can't be communicated.

3. *Question:* Are you provided the equipment and training needed to perform your job safely and successfully?
   Response: Sometimes, because they do not want anything to happen to hinder production.

4. *Question:* How comfortable are the people in your organization in communicating with Management (Supervisor, Middle Management, Senior Manager)?
   Response: We do not often get the opportunity to talk to Managers above our Supervisor. I guess we can talk to them if necessary. We really only trust our immediate Supervisor because we know them.

5. *Question:* How comfortable are you in raising concerns to Management?
   Response: I am comfortable talking to my immediate Supervisor.

6. *Question:* How important is trust to employees in your organization?
   Response: Trust is very important although employees frequently comment on their lack of trust of many Managers. Some employees

have stated that they are not sure that they trust Management because their suggestions or input are not sought or are ignored when provided. They really trust their Supervisor.

7. *Questions:* Are employees encouraged and willing to get involved in solving issues?
Response: Only if Management requests involvement because sometimes they like solving issues and dictating the approach to the workers.

8. *Question:* What words would you use to describe the Management Team?
Response: Absent, interested more in money than quality. Management does not always value the workers who are responsible for the company's success in getting products out and providing customers the products they paid for. Management appears to be self-serving and concerned only about their bonuses.

9. *Question:* How often do you communicate your ideas and suggestions on improving business processes?
Response: Infrequently and only when Management asks.

10. *Question:* How engaged are the employees in your organization when it comes to participating in business outcomes?
Response: Very little or not at all.

9.2.4.6 *Organizational culture individual interviews*

Comprehensive Safety Services conducted interviews with 25 individual workers, where each person was asked to respond to the following questions. The results received from the interview participants were combined and are summarized here.

1. *Question:* What words would you use to describe the culture of your organization?
Response: Very little relationship-building, little involvement of workers, and new ideas are not readily accepted.

2. *Question:* How does communication flow in your organization?
Response: Flows well between my Supervisor and my work teams. Not very well with the senior leadership team.

3. *Question:* Are you provided the equipment and training needed to perform your job safely and successfully?
Response: Mostly. However, there are times when it is a challenge to get what is needed because of cost.

4. *Question:* How comfortable are you in communicating with Management (Supervisor, Middle Management, Senior Manager)?
   Response: I do not often get the opportunity to talk to Managers above are Supervisor. I don't know since I don't communicate with upper Management.

5. *Question:* How comfortable are you in raising concerns to Management?
   Response: I am comfortable talking to my immediate Supervisor.

6. *Question:* How important is trust to you?
   Response: Trust is very important although employees frequently comment on their lack of trust of many managers. Sometimes employees have stated that they are not sure that Management trusts them because their suggestions or input is not sought after or is ignored when provided.

7. *Question:* How would you describe your relationship with your Supervisor?
   Response: Very good. My Supervisor is concerned about their direct reports and ensures that we have the tools we need to get work done safely. My Supervisor is very capable and a knowledgeable leader.

8. *Question:* Are you and the employees on your team encouraged and willing to get involved in solving issues?
   Response: Only if Management requests involvement because sometimes they like solving issues and dictating the approach to the workers.

9. *Question:* What words would you use to describe the Management Team?
   Response: Absent, interested more in the financial bottom line. The Management Team does not show that they value workers, and are concerned only about their bonuses.

9.2.4.7 Organizational culture summary

Evaluation of Spud's Chemical Company showed that the Company has an organizational culture that is not optimal for continual growth and increase. Based on document reviews, surveys, and interviews, Comprehensive Safety Services identified several weaknesses that Spud's Chemical Company should address to improve organizational culture and become more efficient and client-focused. Following are the key company process weaknesses as related to four shared attributes: leadership, employee engagement, organizational factors, and work planning and execution.

9.2.4.7.1 Leadership
- Management is not always good at keeping promises
- Employees generally do not view managers as being good Managers
- Communication is not always timely and accurate
- Management does not involve workers in work decisions

9.2.4.7.2 Employee engagement
- Employees are not actively engaged in the business
- Management does not generally encourage worker involvement
- Opinions of employees are not always valued or sought out

9.2.4.7.3 Organizational factors
- Communication is not fluent and does not flow across the organization
- Employee mostly exhibit low morale
- Roles and responsibilities not always define clearly and understood

9.2.4.7.4 Work planning & execution
- Little involvement of workers in problem solving
- Decisions on work are generally made by Management

9.3 Quantitative evaluation of Spud's Chemical Company

Comprehensive Safety Services utilized a quantitative process to analyze the data gathered from Spud's Chemical Company to develop an improvement plan based on the application of the Peak Performance Model. Each functional element of the model was individually, and then collectively, analyzed to identify weaknesses in Spud's Chemical Company improvement mechanisms, but also management and operational practices. Application of the quantitative process provides a visual means for trending performance and reduces subjectivity and bias that may be inadvertently introduced by the assessment team. By performing a quantitative analysis, information gained from the analysis can then be applied across the Corporation. The quantitative process can be utilized to compare Corporate facilities or even product lines within the same facility. It is a versatile tool that can better pinpoint issues that a company is experiencing and which are preventing the Company from achieving Peak Performance. The quantitative process used, and associated diagrams presented in this book, are proprietary tools developed by the authors.

9.3.1 Lean

A review of Lean for Spud's Chemical Company was performed utilizing quantitative analysis. Lean was appraised using numerical values assigned to specific attributes and organized by the shared attributes: leadership, employee engagement, organizational factors, and work planning and execution; a methodology developed by the authors. Table 9.7 presents a summary of the results. As depicted in Figure 9.1, the overall score was well below 50%, highlighting the lack of focus on any type of continuous improvement processes to reduce waste, streamline processes, and focus on client needs. An overall score below 75% would be considered problematic for an organization. A summary of the issues noted during the assessment that needed attention are listed in the next sections.

Table 9.7 Lean Quantitative Analysis Scores

Attribute	Possible Score	Score
Leadership	30	7
Employee Engagement	30	6
Organizational Factors	15	4
Work Planning and Execution	25	6
Total	100	22

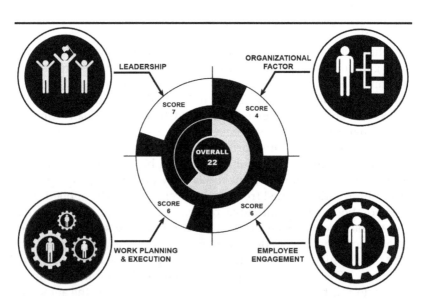

Figure 9.1 Overall Lean score using quantitative analysis. Legend: 0–40 Limited Progress | 41–60 Getting Started Toward Peak Performance | 61–84 Performance Improving | 85–100 Soaring Toward Peak Performance.

9.3.1.1 Leadership
- The Management Team was not outwardly supporting Lean or any other continuous improvement process
- The Management Team was not visible in the plant, observing operations, or looking at the process with a "Lean Lens"
- The Management Team supported a status quo environment

9.3.1.2 Employee engagement
- The Management Team was not easily accessible and/or actively engaged with the workforce
- The Management Team did not foster an environment of supporting employees to improve their jobs or company processes
- Employees were not offered an opportunity to make recommendations to the Management Team on improvements

9.3.1.3 Organizational factors
- The employees have limited access to the Management Team, limiting the opportunities to build trust
- Mission, Vision, and Company goals were not visible to the workforce to enable the entire workforce to support and be involved in the company
- The Management Team did not provide routine communications to the workforce

9.3.1.4 Work planning and execution
- Lean principles were not implemented and incorporated into work planning
- The Management Team had not utilized Lean principles to streamline workflow
- The Management Team had not implemented standard work procedures/practices to ensure repeatability of work execution

9.3.2 Human performance improvement

A review of Human Performance Improvement for Spud's Chemical Company was performed utilizing the quantitative analysis. Human Performance Improvement was appraised employing a worksheet developed by the authors. Human Performance Improvement was appraised using numerical values assigned to specific attributes and was organized by the shared attributes: leadership, employee engagement, organizational factors, and work planning and execution. Table 9.8 presents a summary of the quantitative evaluation. Results of the analysis for Human Performance Improvement, shown in Figure 9.2, indicates the overall score

Table 9.8 Human Performance Improvement Quantitative Analysis Scores

Attribute	Possible Score	Score
Leadership	15	3
Employee Engagement	30	3
Organizational Factors	20	6
Work Planning and Execution	35	11
Total	100	23

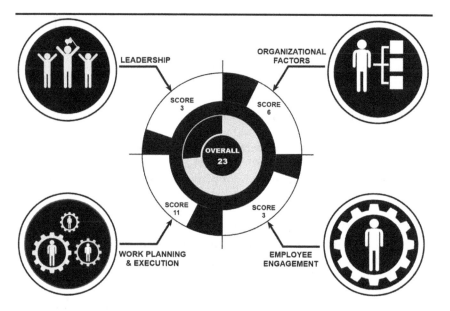

Figure 9.2 Human Performance quantitative score. Legend: 0–40 Limited Progress | 41–60 Getting Started Toward Peak Performance | 61–84 Performance Improving | 85–100 Soaring Toward Peak Performance.

is well below 50%, underscoring the lack of emphasis on reducing human error by understanding processes and taking actions to reduce the potential for the workforce to make errors. An overall score below 75% would be considered problematic for an organization. A summary of the weaknesses identified during the assessment is listed in the following sections.

9.3.2.1 Leadership
- The Management Team failed to develop and issue a strategic plan to reduce human error and improve performance
- The Management Team did not support Human Performance Improvement initiatives

- The Management Team did not interact with the workforce routinely enough to identify potential areas for human error, such as fatigue or distractions with the workforce, etc.
- The Management Team did not offer continuing training to ensure the workforce maintain proficiency
- The Management Team did not track and trend performance to identify areas for improvement

9.3.2.2 Employee engagement
- The Management Team spent limited time in the facility encouraging worker feedback and general communications
- Employees were not coached on the expectations to ensure error-free performance
- Employees were not provided an opportunity to make recommendations

9.3.2.3 Organizational factors
- The Management Team failed to conduct post-job reviews to identify opportunities for improvement
- The Management Team failed to openly address and correct issues
- The Management Team did not set clear expectations

9.2.2.4 Work planning and execution
- The Management Team spent limited time in the plant observing work to understand how work was planning versus how work was actually executed
- The Management Team failed to analyze completed task to look for ways to improve the execution of the activity
- The Management Team did not analyze the process looking for error traps
- The Management Team failed to reinforce the expectation to conduct task previews and pre-job briefs
- The Management Team failed to discuss and reinforce the expectation to identify critical steps

9.3.3 Operations excellence

A review of Operations Excellence at the Spud Chemical Company was performed utilizing quantitative analysis. Operations Excellence was appraised using a worksheet developed by the authors. Operations Excellence was appraised using numerical values assigned to specific attributes and was organized by the shared attributes: leadership, employee engagement, organizational factors, and work planning and execution. Table 9.9 presents the numerical scores associated with each of the attributes for Operations Excellence. Results of evaluating Operations

Table 9.9 Operations Excellence Quantitative Analysis Scores

Attribute	Possible Score	Score
Leadership	25	10
Employee Engagement	30	13
Organizational Factors	30	16
Work Planning and Execution	15	7
Total	100	46

Excellence for Spud's Chemical Company indicated the overall score is close to 50% compared to a perfect score of 100. Quantitative analysis of the data indicated workers are adequately performing their jobs; however, improvements were needed in leadership skills, employee engagement, and culture. Figure 9.3 presents a summary of the results. A summary of the issues noted by Comprehensive Safety Services as needing attention is listed in the following sections.

9.3.3.1 Leadership

- The Management Team needs to improve workforce engagement and communications

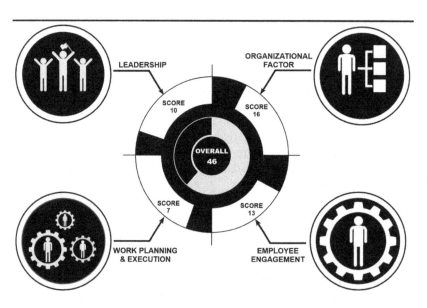

Figure 9.3 Overall Operations Excellence score using quantitative analysis. Legend: 0–40 Limited Progress | 41–60 Getting Started Toward Peak Performance | 61–84 Performance Improving | 85–100 Soaring Toward Peak Performance.

- The Management Team needs to understand the meaning of core values as a driver for managing the plant and the company
- The Management Team does not appear to value organizational learning as a fundamental business tenet

9.3.3.2 Employee engagement
- The workforce appears to not believe they are appreciated; they are lost after Spud's Chemical Company was bought by D.O. Chemical Company
- Attention is needed to ensure company policies and procedures are established for work practices such as overtime, training, and communications
- Time management of workers, and associated operational practices, needs to be evaluated to ensure required communications, and visual evaluation of areas, are performed
- Additional productivity could be achieved by engaging the workforce in feedback of task performance

9.3.3.3 Organizational factors
- There was a significant lack of trust between the workforce and Management
- Management needs to significantly improve organizational and individual communications
- There was a perception that Management does not exhibit a caring attitude

9.3.3.4 Work planning and execution
- A defined process for performing work should be documented and communicated to the workforce
- Worker feedback, with respect to operations, should be solicited and encouraged
- A defined process should be developed for ensuring consistent communication and operational status of equipment

9.3.4 Organizational culture

A review of the culture for Spud's Chemical Company was performed utilizing quantitative analysis. Figure 9.4 depicts a summary of the results. The organizational culture was evaluated using a worksheet developed by the authors. The organizational culture of Spud's Chemical Company was appraised using numerical values assigned to specific attributes and was organized by the shared attributes: leadership, employee engagement, organizational factors, and work planning and execution. As shown in Table 9.10, results of an overall organizational culture score below 50%

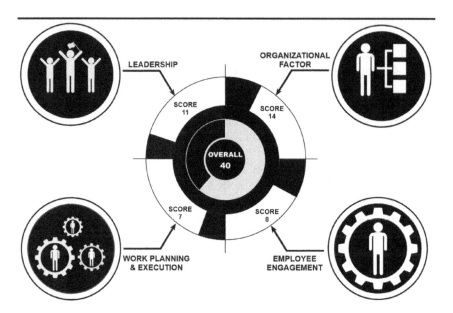

Figure 9.4 Overall organizational culture score using quantitative analysis. Legend: 0–40 Limited Progress | 41–60 Getting Started Toward Peak Performance | 61–84 Performance Improving | 85–100 Soaring Toward Peak Performance.

Table 9.10 Organizational Culture Quantitative Analysis Scores

Attribute	Possible Score	Score
Leadership	30	11
Employee Engagement	20	8
Organizational Factors	40	14
Work Planning and Execution	10	7
Total Score	100	40

indicate that the culture of the organization should become a focus area for Management to achieve improvements in operational performance. A culture score below 75% would be considered problematic for an organization. A summary of the issues noted that should be focused on in each of the areas are listed in the following sections.

9.3.4.1 Leadership

- Management support is not always shown.
- Vision and goals are not always set and clear.
- Communication is not generally timely, open and honest, or clear.
- Employees believe that their opinions are not generally sought or their concerns listen to by Management.

- Leaders are not viewed as accountable or respectful of workers.
- Employees generally do not trust Management beyond their immediate Supervisor.

9.3.4.2 Employee engagement
- Employees do not generally feel connected to their organization or free to raise issues.
- The opinions of employees are not generally sought out by Management.
- Workers do not always know and understand performance expectations and are not generally willing to offer suggestions freely.

9.3.4.3 Organizational factors
- The workers resist change or are slow to accept change.
- The vision of the Company is not clear to all.
- Decisions are made by Management and employees are not generally consulted.
- Teamwork is apparent among a small segment of teams and not openly embraced.
- Policies and procedures are not always followed.

9.3.4.4 Work planning and execution
- Employees are at times not included in the work planning process although they are responsible for work execution.
- A higher degree of focus on safety when planning work is needed.

9.3.2.5 Cumulative quantitative analysis
The real secret to achieving Peak Performance is leadership engagement and trust among workers. The leadership team must be in alignment to enable significant changes to occur within the organization. The employees must trust that the leadership is looking out for them and has their best interest at heart for any significant change. If everyone in the organization understands the direction the Company is taking and *why* the Company wants or needs to make changes, the workforce will be more accepting of the changes and more willing to make the effort necessary to change. The other important key is employee engagement, talking with employees, helping them understand the impacts, but also asking employees for their input, what will make their jobs easier, more streamlined, less complex. Employees must feel valued, needed, and appreciated if you want to get 100% all day every day. Table 9.11 depicts the Peak Performance Model analysis scores for Lean, Human Performance Improvement, Operations Excellence, and Organizational Culture as well as the total score of 33 for Peak Performance.

The following recommendations were developed by Comprehensive Safety Services (after an extensive review) for the Spud Chemical Company.

Table 9.11 Peak Performance Quantitative Analysis Scores

Attribute	Possible Score	Score
Lean	25	6
HPI	25	5
Operations Excellent	25	12
Organizational Culture	25	10
Total Score	100	33

These recommendations will start the journey to Peak Performance for the Spud Chemical Company. The recommendations were organized by the four primary attributes: leadership, employee engagement, organizational factors, and work planning and execution.

Figure 9.5 shows the individual quantitative scores for each functional element, as well as the total score for the Peak Performance Model, utilizing the worksheet developed by the authors. When integrating the scores for each element of the model only 25% of each of the individual scores were inserted because the model assumes that each of the four elements are equally weighted. Therefore 25% of the scores received for each element forms the overall appraisal score. Again, the review highlighted

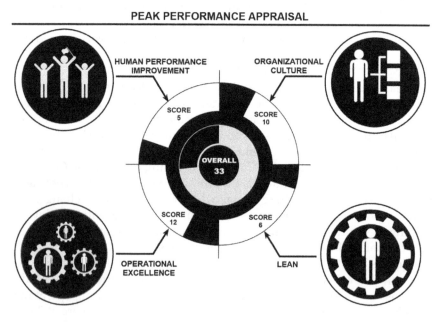

Figure 9.5 Overall Peak Performance score using quantitative analysis. Legend: 0–40 Limited Progress | 41–60 Getting Started Toward Peak Performance | 61–84 Performance Improving | 85–100 Soaring Toward Peak Performance.

the need for significant improvement if Spud's Chemical Company was going to reach and maintain Peak Performance. The D.O. Chemical Company learned they could apply the lessons learned from the event at Spud's Chemical Company to other companies they owned. The D.O. Chemical Company also used Comprehensive Safety Services to evaluate all their similar operations to establish a baseline for measuring quantitative improvement in Corporate business practices and operations. The consolidated recommendations that would be incorporated into an operational improvement plan and drive Spud's Chemical Company to achieve peak performance are listed in the following sections.

9.3.2.5.1 Leadership
- Establish company expectations for performance, which includes identifying mission and business critical processes
- Develop a management strategic plan to drive continuous improvement
- Establish defined roles and responsibilities, shared vision, and communicate expectations for operations to the workforce

9.3.2.5.2 Employee engagement
- Spud's Chemical Company needs to establish frequent interactions between management and the workforce. The focus of the interactions is to encourage feedback and input, and build trust. Example forums that promote such interactions include roundtable meetings, management field observations, specific union and management meetings, safety councils, and employee recognition programs.
- Establish routine provide written and oral communications. Examples of such routine communications include daily shift briefs, informal dialogue during shift changes, monthly newsletters, monthly videos from the Company.

9.3.2.5.3 Organizational factors
- Establish a transparent and responsive issue-resolution process. This could be done with a website that lists the questions asked and the responses.
- Reinforce worker involvement in the issue-resolution process. The company could utilize Lean principles to support issues resolution and involve the workers on the teams.
- Establish and communicate company goals and objectives that are visible to, and embraced by, the workforce. The status must be tracked and progress visible to the workforce. This can be accomplished by multiple means such as posters on the walls, publishing monthly metrics on a website, etc.

9.3.2.5.4 Work planning and execution
- Incorporate and monitor worker involvement in the work/task planning process. This could start with a walkdown of the task to be performed, followed by asking lessons learned from the last time the task was performed, soliciting their input as to how they would perform the work, etc.
- Incorporate error-reduction tools into company and Corporate work instructions.

9.4 Spud's Chemical Company Operational Improvement Plan

With the targeted areas of weaknesses identified, Comprehensive Safety Services developed an improvement plan that focused on strengthening and sustaining improved operational and safety performance. The improvement plan could have been developed from results generated by applying the qualitative or quantitative approach of the Peak Performance Model; however, Comprehensive Safety Services applied the quantitative model so the D.O. Chemical Company could further use the Peak Performance Model across their operational product lines to achieve consistent, improved corporate performance. Table 9.12 is an excerpt from Spud's Chemical Company Operational Improvement Plan.

Comprehensive Safety Services recommended Spud's Chemical Company communicate why the assessment was requested, the overall results, and the path forward to all employees and subcontractors. They also recommended that the workforce be involved in acceptance and implementation of an improvement plan, along with communications on progress toward the goals through the use of performance metrics that were of meaning to both Management and the workforce. After these recommendations were fully implemented, the D.O. Chemical Company requested Comprehensive Safety Services perform a follow-up review to evaluate effectiveness of implementing the recommendations. This iterative review process will support the Spud Chemical Company's journey to peak performance. It may take several iterations of assessing, reevaluating actions, and executing the plan, but the end result will be the achievement of Peak Performance.

As depicted for Spud's Chemical Company, the tools provided in this book will enable any company to complete a comprehensive assessment to identify key areas on which to focus improvements and strive to achieve Peak Performance.

Chapter nine: Spud's Chemical Company, LLC

Table 9.12 Excerpt from Spud's Chemical Company Operational Improvement Plan

Area of Focus	Activity	Targeted Completion Date	Responsible Organization	Performance Measure
Defined company expectations for performance, which includes identifying mission, vision, core values, and business critical processes	Develop a Management Strategic Plan for Spud's Chemical Company	Within 60 days of completed Comprehensive Safety Services Assessment	Senior Management	Schedule – publishing of Document and Communication to Workforce
Established roles and responsibilities of Management and the workforce	Review company policies and procedures and identify gaps in defined roles and responsibilities	Within 30 days of completed Comprehensive Safety Services Assessment	Mid-Level Management and Human Resources	Performance metric of time vs. number of documents reviewed
	Revise policies and procedures to reflect defined roles and responsibilities	60 days after completion of the gap analysis	Assigned Management	Schedule performance metric
Established interactions between Management and the workforce. The focus of the interactions is to encourage feedback, input, and build trust	Establish and schedule Management Roundtable Meetings, routine Union and Management Meetings, Employee-Led and Executive Management Sponsored Safety Committee	30 days after issuance of Comprehensive Safety Services Assessment analyze and establish meetings	Senior Management and Human Resources	Schedule metric communicated at all meetings
	Establish and implement a monthly communication letter	60 days after completion of the Comprehensive Safety Services Assessment	Employee-Led Team	Schedule metric
	Establish an Employee Recognition Program driven by worker feedback	60 days after completion of the Comprehensive Safety Services Assessment	Senior Management and Employee Team	Performance metric that depicts number of recognitions over time

(Continued)

Table 9.12 (Continued) Excerpt from Spud's Chemical Company Operational Improvement Plan

Area of Focus	Activity	Targeted Completion Date	Responsible Organization	Performance Measure
Transparent and responsive issue-resolution process. This could be done with a website that lists the questions asked and the responses	Develop an issue resolution process that is responsive to worker suggestions, as well as deficient operational conditions. The company should consider incorporating Lean principles to support resolution of issues	180 days after completion of the Comprehensive Safety Services Assessment. The selected issue resolution process must be compatible with any systems used by Corporate	Senior Management and Employee Team	Schedule metric for acquiring system. Performance metrics that depict number of issues submitted, time required to resolve the issues, and worker feedback of the process
Documented work control process	Review current work operational processes to identify and document a consistent approach for work planning. This activity should also address deficiencies associated with the Maintenance program	180 days after completion of the Comprehensive Safety Services Assessment. The selected work planning process should be consistent with Corporate objectives	Operational Management and Employee Team	Schedule metric to manage achieving goals. Performance metrics that monitor how well key parts of the work planning and control process is executed

(Continued)

Table 9.12 (Continued) Excerpt from Spud's Chemical Company Operational Improvement Plan

Area of Focus	Activity	Targeted Completion Date	Responsible Organization	Performance Measure
Worker involvement in the work/task planning and control process	Incorporate into work planning and control process worker walkdowns, and mechanism that is transparent for worker involvement in the planning and control process	180 days after completion of the Comprehensive Safety Services Assessment. This activity is to be completed concurrently, or as part of the effort to develop and document a consistent work planning and control process	Operational Management and Employee Team	Performance metric that targets the areas of the work planning and control process where worker involvement is critical
Incorporate error reduction tools into Company and Corporate work instructions	Identify human performance activities and error reduction tools to be applied when operating. This activity should be conducted concurrent with the development of the defined work planning and control process	60 days after completion of the documented work planning and control process and review of company policies and procedures	Operational Management and Employee Team	Performance metric that targets application and incorporation of key error reduction techniques in the work planning and control process. This should also address critical information needed as part of shift turnover

Index

A

A3, 26, 28
Active error, 43, 49
Administrative use procedures, 74
Anatomy of an event, 44–50
Assessment, 120
Attributes, 2, 3, 4

B

Benchmarking, 94–95, 100
Box plot, 31, 33

C

Change management, 102
Communication, 133, 137, 139, 145
Comprehensive safety services, 166
Comprehensive strategy, 1
Confirming assumption, 78–79
Contractor oversight, 93–94
Control chart, 37, 38
Core function, 110, 111–123, 126–128
Critical step, 65, 66, 67, 69, 72, 84, 103–108
Culture, 129, 130, 132, 133, 134, 135, 137, 138, 139, 144, 145, 146, 147, 148, 149, 150, 151, 153, 156, 157, 158

D

Decision making, 91–93
Defective controls, 49–50
Distributing responsibility, 82

E

Employee competencies, 44, 46–47, 48
Employee engagement, 147, 148, 154, 155, 157

Error precursors, 44–48, 49
Excess motion, 17, 19–21

F

Facilitator, 112
Fishbone, 90, 91
Fishbone diagram, 31, 33, 34
5-Why, 31, 34
Flagging, 85–86
Focus groups, 30, 31, 32

G

Group thinking, 83

H

Halo effect, 81, 82
Human capital, 17, 23–24
Human characteristics, 44, 48
Human performance, 4, 6–7, 13, 41–63, 147, 148, 150, 158, 169, 181, 197, 198, 203, 209
Human performance tools, 65, 66, 67, 104, 106

I

Independent checking, 84–85
Independent oversight, 99–101
Initiating event, 48–49
Interviews, 30, 32
Inventory, 17, 19, 20, 21, 26

J

Job site review, 67–68

211

K

Knowledge-based performance mode, 58–61, 70, 90, 91, 107

L

Latent errors, 43–44, 50
Lean, 2, 4–6, 13, 14, 15, 16, 17, 24, 25, 26, 38, 148, 149–150, 169, 174, 175, 195–197
Likert, R., 139

M

Mission critical, 109, 110, 111–113, 114, 115, 118, 119, 120, 121, 123, 126, 127
Monitoring plan, 37

O

Observations, 30, 31, 33
Operational recovery plan, 168
Operations excellence, 4, 7–8, 13, 109–128, 147, 148, 150–151, 169, 182–188, 199–201
Organizational culture, 8, 9, 13, 50–51, 61, 169, 188–195, 201–206
Organizational performance, 9, 10, 13
Overproduction, 17, 21–22, 23

P

Peak performance, 147–159
Peak performance model, 4–10, 13–14, 110, 121
Performance modes, 41–63
Phonetic alphabet, 76–78
Pilot/co-pilot, 81
Place keeping, 79
Plan-do-check-act (PDCA), 34–35
Pre-inspection checklist, 164
Productivity, 147, 148, 150

Program elements, 110, 113–115, 117, 118, 119, 120, 121, 122, 124, 125, 126, 127, 128

Q

Qualitative evaluation, 169–195
Quantitative evaluation, 195–206

R

Record reviews, 33
Response plan, 37
Rework, 17, 22–23
Rule-based performance mode, 54–58

S

Schein, E. H., 129
Shift brief, 164
Skill-based performance mode, 52–54
Strategic management, 2
Strategic plan, 112
Supplier Input Process Output Customer (SIPOC), 26, 28, 29
Survey, 30, 32, 120
Swim-lane, 34, 35, 36

T

Task requirements, 44, 46
Travel time, 17–18

V

Value stream mapping, 26

W

Waiting periods, 17, 21
Weighted decision matrix, 34, 35–37
Whistleblower, 130–131